M000281681

STRATEGY
IN THE
DIGITAL
AGE

MASTERING DIGITAL
TRANSFORMATION

MICHAEL LENOX

STANFORD BUSINESS BOOKS
Stanford, California

Stanford University Press
Stanford, California

© 2023 by Michael James Hale Lenox. All rights reserved.

No part of this book may be reproduced or transmitted in any form or by
any means, electronic or mechanical, including photocopying and recording,
or in any information storage or retrieval system without the prior written
permission of Stanford University Press.

Special discounts for bulk quantities of Stanford Business Books are available
to corporations, professional associations, and other organizations. For
details and discount information, contact the special sales department of
Stanford University Press. Tel: (650) 725-0820, Fax: (650) 725-3457

Printed in the United States of America on acid-free, archival-quality paper

Cataloging-in-Publication Data available upon request.
Library of Congress Control Number: 2022037728
ISBN: 9781503635197 (cloth), 9781503635760 (ebook)

Cover designer: Kevin Barrett Kane
Cover illustration: Adobe Stock

To my children, Ben and Haley, digital natives who are navigating this brave new age

Contents

Figures

Preface

The digital age is upon us. Digital technology is having an impact on virtually every endeavor in our lives. From the constant presence of our mobile devices to the ubiquity of social media to digitally enabled everyday objects such as our cars, appliances, and watches, the digital age is inescapable. Each transaction and engagement in this digital world generates massive amounts of data that when combined with advanced analytics such as machine learning and artificial intelligence manifest opportunities for creating value for stakeholders in new and transformational ways. They also raise numerous ethical and legal concerns that cannot be avoided and require purposeful attention.

To say that navigating the digital age is fraught with peril is an understatement. I have not found an industry or sector that is not feeling the pressure to digitally transform. Most companies and organizations that I talk and work with are afraid of being digitally disrupted. They are faced with evolving or going extinct. Big tech companies have built powerful platforms that they are leveraging to enter and, in many cases, dominate adjacent sectors. No sector is immune. Finance, energy, manufacturing, media, consumer goods, agriculture, all feel the pressure to transform. The criticality of data and analytics is fundamentally reshaping markets and leading to winner-take-all dynamics in many industries. The list of venerable companies left behind in the wake of digital disruption grows every day.

Strategy in the Digital Age is designed to help businesses and organizations navigate this brave new world. First and foremost, this book is aimed at those undertaking a digital transformation at their organization or who are simply looking to be more savvy strategic thinkers in the digital age. To be clear, this book is about strategy. We do not take deep dives explaining the intricacies of machine learning algorithms or how to set up a blockchain. This book is not aimed at IT professionals looking for the specifics of building a data lake or migrating to the cloud. There are numerous resources out there that can provide detailed guidance on these topics. This book is about looking at the big picture.

A common theme throughout this book is that digital technology is changing the underlying basis of competition for a wide swath of businesses and industries. Digital transformation is much more than building a digital infrastructure to gather and process data (though this remains critical). It is about understanding how digital technologies enable the creation of innovative value-added services and products. It is about how new competitive positions and business models are possible in the digital age and how to think critically about how to both create and capture value. Most important, this book is about how to lead a digital transformation in your organization—not only paying attention to the technical details but also thinking through the human dimension and being sophisticated about the numerous social and policy challenges raised by digital technology.

Strategy in the Digital Age is based on my second-year MBA elective of the same name that I have been teaching for several years at the University of Virginia's Darden School of Business. I am incredibly grateful to my students who have served as "beta testers" for this material and who have provided extremely valuable feedback that has greatly improved the final project. I also want to thank my partners at the Boston Consulting Group—Amane Dannouni, Ching Fong Ong, and Sonja Rueger—whom I collaborated with on an online course from Coursera titled "Digital Transformation." I want to thank the numerous executives and companies that I have worked with, either as a consultant or through Darden's Executive Education and Life-Long Learning operation. Your willingness to share both your successes and struggles has been critical to my thinking on the topic.

In undertaking this project, I also benefited greatly from the writings of and my interactions with my colleagues at the Darden School and beyond. I have been studying, researching, and teaching about technology and disruption for over a quarter century. I thank my previous coauthors, Gary Dushnitsky at London Business School, Chuck Eesley at Stanford University, Andy King at Boston University, and Scott Rockart at Duke University, who have profoundly shaped my thinking on disruption. In my research for this book, the recent works of Nicolaj Siggelkow and Christian Terwiesch at the Wharton School of Business, Marco Iansiti and Karim Lakhani at the Harvard Business School, Ajay Agrawal, Joshua Gans, and Avi Goldfarb at the University of Toronto, and Ron Adner at Dartmouth University, among many others, have contributed to my own thinking on the topic. I have benefited greatly from my wonderful colleagues at the Darden School of Business including, but not limited to, Alex Cowan, Ed Freeman, Yael Grushna-Cockayne, Jared Harris, Ed Hess, Jeanne Liedtka, Bobby Parmar, Saras Sarasvathy, Scott Snell, Raj Venkatesan, and Venkat Venketaraman.

This book would not have been possible without the assistance of my coauthor and colleague Becky Duff, who graciously agreed to help with fact checking and citations. The figures in this book were designed by Leigh Ayers. Thanks to Steve Catalano and the entire team at Stanford University Press for their enthusiasm and guidance. I also thank my external reviewers. I greatly appreciate your suggestions and feedback. I would like to acknowledge the support of the Darden School and the Batten Institute for Innovation and Entrepreneurship. A special thanks to Kush Arora, Darden's Chief Digital Officer, who has been helping lead Darden's digital transformation efforts and has taught me so much about data lakes, business intelligence software, dashboards, and, most important, how to effectively manage a transformation.

Last, I wish to thank my family. My two children were born soon after the dot-com era. Google, Facebook, Uber, Airbnb, Tesla, TikTok, and hundreds of other digital disruptors have risen during their lifetimes. As they have grown up, we have had to navigate mobile devices, social media, streaming media, and online learning. Questions of screen time, online presence, data privacy, cyberbullying, and cybersecurity are constant conversations. The digital age is stressful, especially for parents. I am thankful to have been traversing these disruptive times with my wife and kids. Their support, insights, and desire to be good digital citizens inform and inspire me every day.

STRATEGY IN THE DIGITAL AGE

1

Rise of the Digital Age

On January 19, 2012, Kodak declared bankruptcy. On the surface, this was entirely unsurprising. Kodak was the global leader in the manufacturing and processing of chemical-based photographic film. One of the iconic companies of the twentieth century, Kodak was brought to its knees by the rise of digital imaging and the reconfiguration of the value chain for pictures. Seemingly overnight, we stopped taking pictures with film-based cameras, printing them out for sharing in photo albums and picture frames, and started using our smartphones from the likes of Apple and Samsung to create digital images shared through social media sites such as Facebook and Instagram.

Except the change did not occur overnight and Kodak was not asleep at the wheel. In fact, as early as the late 1970s, Kodak recognized the disruptive potential of digital imaging and began investing heavily in its own digital transformation. Kodak spent billions on R&D. It had some of the first patents for digital image sensors. It introduced one of the first digital cameras, in 1989.[1] It created a new architecture for sharing photos, pioneering the PhotoCD and CD player in the early 1990s. It hired a tech-savvy CEO away from Motorola in 1993 and was a leader in point-and-shoot digital cameras for a time. This wasn't enough.

Kodak is not alone. The pressure to digitally transform knows no bounds, having an impact on virtually every industry and organization. The list of digital casualties is long and growing every day. Blockbuster,

Borders, Nokia, Sears, and Blackberry were all raging successes laid low by digital transformation, or as I like to call digital transformation's "evil twin," digital disruption. In their wake has risen a group of innovative companies with an entirely new wiring: Apple, Amazon, Google, Facebook, Uber, Airbnb, Netflix, Box, Zoom, and assuredly a whole host of upstart ventures that have not yet become household names. These are companies that have been built digitally from the ground up, that understand the primacy of data and how to leverage data to add value to their offerings.

While there is a temptation to view digital transformation as new, it is important to recognize that digital transformation has been occurring for well over a half century. In 1972, Hamilton introduced the first digital watch, the Pulsar.[2] Powered by a battery-and-quartz technology, the digital watch was a major disruption to the centuries-old analog watch industry. Suddenly, generations' worth of expertise in making small springs and gears was rendered moot. A decade earlier, the electronic typewriter pioneered by IBM disrupted the centuries-old mechanical typewriter business and established leaders like Remington—only to be disrupted a decade later with the rise of personal computing and word processing.

Yet there are reasons to believe that digital transformation is accelerating. The emerging digital infrastructure is giving rise to enormous datasets, what we used to call "big data," that are helping fulfill the promise of artificial intelligence and machine learning. These are creating virtuous cycles in which data and learning beget more data and learning, accelerating innovation and creating the risk that those slow to digitally transform will be quickly left behind. I use the analogy of building an onramp to a highway. Once you successfully complete the ramp and enter the highway, you are speeding down the digital highway. Digital transformation is about building that onramp to the highway. If you are too slow, you may never catch up to those who came before you.

Over 50 percent of Fortune 500 companies have either been acquired, merged, or declared bankruptcy since 2000.[3] In recent years, stalwart companies such as AT&T and ExxonMobil have been removed from the Dow Jones Industrial Index to be replaced by tech companies, Apple and Salesforce, respectively. The average lifespan of S&P 500 firms has been steadily declining, creating more turnover among the largest firms.[4] Serial entrepreneur Tom Siebel, founder of artificial intelligence company C3 (ticker AI on the NYSE), points out that "it is estimated that 70 percent of the companies in existence today will shutter their operation in the next 10 years. . . . Mass extinction events don't just happen for no reason—I believe the causal factor is digital transformation."[5]

This is not to say that digital transformation needs to be fatal. In the photographic film industry, Kodak rival Fuji doubled down on its core chemistry expertise and diversified into several adjacent industries such as health care, optics, and chemicals. Similarly, film-based camera manufacturer Nikon pivoted quickly to higher-end consumer digital cameras, leveraging its expertise in lenses to differentiate from smartphone alternatives. IBM has transformed itself at least twice: from the maker of mechanical business machines to the global leader in mainframe computers to its current manifestation as a digital solutions provider leveraging Watson, its AI technology platform.

This book is about how to understand the strategic ramifications of the digital age and to design and execute strategies to help your organization flourish. In my work with business leaders, I have yet to find a business or industry that is not feeling the pressure to digitally transform. All too often, companies and organizations relegate their digital transformation efforts to the IT team, as if wrangling your data and creating a few nice data visuals is all it takes. If only it was so simple! This book looks at how digitization is transforming the very ways organizations deliver and capture value, creating new business models, disrupting existing value chains, and providing opportunities for building new enduring sources of competitive advantage for those who have the foresight and tenacity to capture them in this digital age.

THE EXPONENTIAL GROWTH OF THREE CORE TECHNOLOGIES

The growing technology reckoning wrought by digitization has had a long gestation. The rise of the digital age has been catalyzed by the exponential growth in three core digital technologies: processing power, storage capacity, and bandwidth (see Figure 1.1). In 1965, Intel founder Gordon Moore advanced his now famous observation that the number of transistors per integrated circuit tends to double every year (later he revised it to roughly every two years). Referred to as Moore's Law, his observation has proven remarkably prescient. Over the past fifty years, the increased density of transistors on integrated circuits has allowed microprocessors to exponentially increase in processing power while massively lowering their cost per gigahertz. Stated another way, we have seen a one-trillion-fold increase in floating point operations per second of computer processing power. An Apple Watch on your wrist today has nearly double the processing power of the Cray-2 supercomputer from 1985.

Similar dynamics have been observed in storage and bandwidth. Magnetic storage, in terms of bits per dollar, has progressed from roughly 1 bit per dollar in the early 1950s[6] to 350 billion bits per dollar in 2020. From 5-1/4-inch floppy disks with 360 kilobyte capacity in the 1980s to USB flash drives with 8 megabyte capacity in the early 2000s to the petabytes of capacity available through cloud services today, we have seen an exponential increase in capacity per dollar for nearly seventy years. Similarly, Internet bandwidth has grown exponentially from 1,000 bits per second using dial-up modems in the early 1980s[7] to 1 billion bits per second in 5G networks. We think nothing of streaming videos on our phones today when just twenty years ago downloading a simple picture using an AOL dial-up service could feel interminable.

Such exponential growth is hard for us to comprehend, let alone design strategies to accommodate. To help illustrate, let me evoke the apocryphal story of the creator of the classic game of chess. As the story goes, the inventor of chess was an Indian man who was summoned by the ruler of India to his court to celebrate his signature creation. The ruler, wishing to honor him, asked him to suggest a reward for his invention. The inventor, being a smart man, asked for one grain of wheat to be placed on the first square of the chess board and then to double it on the next square and then double it again and again until all 64 squares on the board were covered. The ruler scoffed at what seemed like a pittance. Assuredly, the inventor deserved more than a modest pile of wheat. What the ruler didn't appreciate is that doubling 64 times, in other words 2^{64}, results in over 18 quintillion grains—about two thousand times the global production of wheat in 2020.

This is the power of exponential growth. While there is no consensus on how long Moore's Law or the exponential growth in bandwidth and storage will continue, what is clear is that even a few more years of doubling could lead to a radically different world than the one in which we

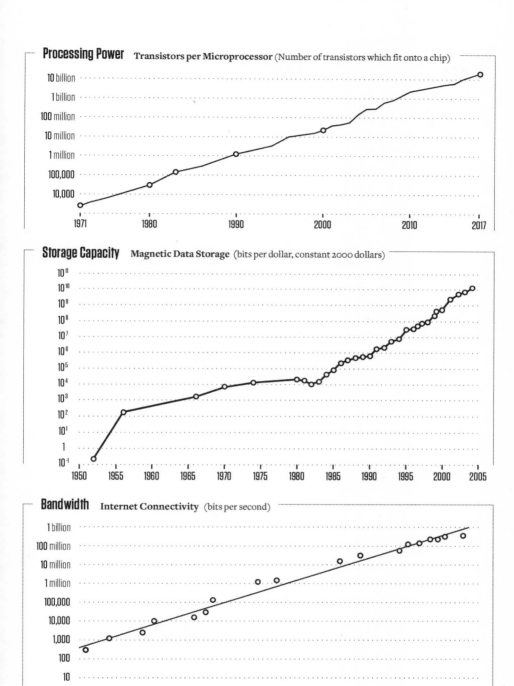

Figure 1.1. Exponential Growth of Technology. Source: Karl Rupp and Nielsen Norman Group, https://www.nngroup.com/articles/law-of-bandwidth/.

currently live. Technical challenges that may sound daunting today may be trivial in a decade. This, at its heart, is the challenge of the digital age. We live in a fast-moving world where the start-of-the-art technology of today could be obsolete by next year. The implications for competition are huge and highlight, yet again, the importance of clear thinking and well-devised strategies.

THE PRIMACY OF DATA

So, what are we doing with all this processing power, storage capacity, and bandwidth? In the simplest terms, the exponential growth of these three core digital technologies is ultimately driving an explosive growth in data. By the end of 2020, there were approximately 60 trillion gigabytes of data (40 zettabytes) in the world. That is up from 1.2 zettabytes accumulated by 2010.[8] Or put another way: 90 percent of all the data in the world was generated in the last two years. What was considered "big data" just a few years ago pales in comparison to what we create daily these days. For example, every day, Google gets over 3.5 billion search requests and WhatsApp users exchange up to 65 billion messages.[9]

Several technological developments are facilitating this massive collection and sharing of data. Cloud computing is providing a reliable, secure source of storage in huge volumes at low costs. Cloud computing is very simply the use of remote data centers—stacks and stacks of servers—that provide on-demand access via the Internet to data storage and computing power without direct active management by the user. Over the past decade, we've seen a massive shift from local servers to the cloud. By the end of 2020, roughly two-thirds of enterprise infrastructure resided in the cloud.[10] Amazon led cloud services providers with a 32 percent market share in 2020.[11] Increasingly, the cloud is a place for storing not only files but natively running applications. Microsoft 365 leverages Microsoft's Azure cloud services platform to run their famous Office Suite.

One should note that cloud computing is feasible because of the gains made not only in storage, but in bandwidth and processing power as well. This is also true of the rise of the Internet of Things (IoT). IoT refers to the hyperconnectivity of objects with the Internet. As of 2021, there were over 46 billion devices globally connected to the Internet.[12] In the United States, the average number of connected devices per household in 2020 was 10—largely driven by mobile and personal computing and consumer electronics such as watches, televisions, and gaming systems.[13] By 2030, the number of connected global devices is expected to grow to 125 billion[14] as more and more devices are connected—everything from consumer goods such as refrigerators and washing machines to Peloton exercise bikes to commercial equipment such as power generators and industrial robots. Engineers are even placing connected sensors into concrete structures such as bridges and roads to be able to continually monitor their efficacy and safety.

Each of these objects creates data, often massive amounts. A simple Nest home security video camera can generate up to 400GB of data a month.[15] Tesla collects upward of 4GB from a single car per day as users drive around with cameras throughout the car observing their surroundings.[16] With speeds upward of 1.5 gigabits per second, 5G cellular networks, in particular, are opening up the opportunity for real-time exchange of data with both stationary objects such as buildings and mobile objects such as autonomous vehicles.[17] This has huge implications for IoT and will drive greater digitization, creating smart cars, smart factories, smart grids, and smart cities—each generating terabytes of data that can be connected and shared with other data.[18]

Blockchain is another technology that could facilitate massive consolidation and sharing of data. Over-hyped and often misunderstood, blockchain simply is a distributed ledger—think of a spreadsheet or database that can be easily shared on a network. What distinguishes a blockchain from, say, a shared Google doc is that anyone can see the data, but they cannot corrupt or falsify it because every entry (transaction) in this ledger is authorized by a digital signature of the owner. For example, Maersk and IBM have developed a blockchain solution for the global shipping industry that allows all parties along the supply chain to have a secure, reliable, and transparent way to share details as a product makes its way from a producer to an end user or retailer. Suddenly, what was a series of data bottlenecks at customs offices and ports-of-entry was reduced, allowing data to be connected and analyzed—promising to improve efficiency and decrease delays.

Like cloud computing and IoT, blockchain has been enabled by increases in processing, storage, and bandwidth. The most popular blockchain application, the cybercurrency Bitcoin, requires approximately 120 quintillion calculations per second (exahashes or hashrate in blockchain speak) to "mine" new bitcoins.[19] To give a sense of the processing power required, the electricity needed for that amount of computing power is equivalent to 120 gigawatts, or 49,440 wind turbines generating power at peak production per second.[20] Those exahashes are critical to the functioning of the blockchain, as they are what allow for cryptography, or encryption, to secure data. Hash encryption stores relevant information in a new block that is transmitted across the world and added to a blockchain after verification.[21]

Blockchain technology may be useful for a variety of applications in which sensitive data needs to be shared. Cybercurrencies such as Bitcoin can help facilitate payments in countries where third-party intermediaries such as banks are unreliable or corrupt. Blockchain holds great promise in health care to allow for exchange of sensitive health data. Blockchain-driven smart contacts can automate contracts in an IoT world where transactions between entities could increase exponentially. All and all, we live in a world awash in data that will continue to grow exponentially as cloud computing, the Internet of Things, and blockchain, among other technologies, allow for more data collection and sharing.

THE POWER OF AGGREGATION

Why is all this data so important? Consider the music industry. In the 1980s, we saw the rise of the digitization of music as analog formats such as records and cassettes gave way to the compact disc. Suddenly, a vast catalogue of music was converted to "1's" and "0's"—easing replication and sharing. As the Internet and web browsers proliferated in the 1990s, opportunities for exchanging music on massive networks increased. In June 1999, Napster released its software, greatly accelerating connectivity and sharing of music. In April 2003, Apple capitalized on the trend (and the legal woes of Napster) and launched the iTunes Music Store, catalyzing the age of downloads.

This age was short-lived, however. In November 2004, Pandora launched its Internet Radio service, providing recommended playlists based on user data. In October 2008, Spotify followed with its streaming music service, giving users even more freedom to choose their music. In a relatively short period of time, downloads decreased and subscription and ad-based streaming services became the dominant mode of consuming music. Central to the streaming services was leveraging the vast data collected on users' listening preferences to present curated recommendations while also generating data valuable to advertisers looking to target specific customers.

In the music industry, we saw a relatively quick evolution from digitization to sharing files on the Internet to leveraging data to create new business models (see Figure 1.2). Or, in the words of Adner, Puranam, and Zhu, from representation to connectivity to aggregation.[22] Representation refers to the creation of digital artifacts—transforming what might be analog data into a form that is easily replicable and sharable. Connectivity refers to the ability to share digital assets on various platforms, typically through the Internet. Aggregation refers to the ability to leverage, or aggregate, data to create value for stakeholders.

This last phase highlights the critical importance of analytics—the systematic analysis of data. The data captured by streaming services is incredibly valuable. It allows service providers to enhance their value proposition for consumers (for example, curating lists and making recommendations). Equally important, analytics allowed for the emergence of new business models including "freemium" approaches that provide advertising-supported options for free to consumers and subscription-based upgrades to have those annoying ads removed. Data and analytics fundamentally changed the nature of competitive advantage, deconstructed the traditional value chain—to the great consternation of musicians and record labels—and led to an influx of upstart entrants. This is a recurring pattern that we talk about in detail in Chapter 3.

Analytics is ultimately about making better predictions. What music would a person like to hear? What ads are most effective and for what products? What impact will a change in a production process have on overall efficiency? Data and analytics are nothing new. Businesses and organizations have been capturing and analyzing data since the dawn of markets. Every MBA builds their statistics toolkits, learning techniques, and software to facilitate analysis, from regression models to Excel spreadsheets to programming in R. They do this in the service of making better predictions and informing action. To the extent things have changed, it is the scale of the data available and the power of the analytics at one's disposal.

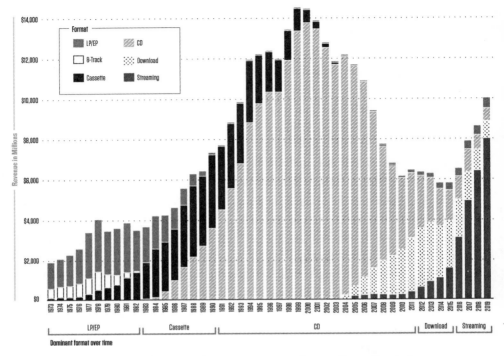

Figure 1.2. Music Industry Evolution, 1973–2019. Source: Recording Industry Association of America, https://www.riaa.com/u-s-sales-database/.

This is particularly true of techniques to help analyze massive datasets in real time. There is so much data being created that often no individual can get their mind around the data, let alone generate interesting insights from it. Data wrangling—merely capturing and organizing the data at an organization's disposal—can be incredibly challenging even for the best-run of organizations. The creation of dashboards and compelling visuals, often updated continuously in real time as data is generated, has become foundational to any digital transformation effort. But to be clear, this is digital transformation at its most basic. True transformation relies on the ability to create compelling predictions based on rigorous analytics applied to the mountains of data at your disposal.

So, what are some of the techniques that allow you to analyze the mountains of data available? Artificial intelligence (AI) probably first comes to mind for many people, and, more specifically, machine learning approaches to building predictive algorithms. Consider Spotify. It has too many customers and too much data to cultivate personalized playlists "by hand." It does not have the resources to support individual analysts crunching data in a spreadsheet to generate a forecast. Rather, it has automated the process using machine learning for populating features such as its Discover Weekly recommendation engine. First, it relies on an approach called collaborative filtering with which it makes predictions based on the behavior of similar users. Second, it leverages natural language processing to analyze blog posts, news articles, and social media posts to identify artists and new trends that cluster into certain categories. Third, it uses convolutional neural networks to analyze the music itself to identify similar traits between songs.[23]

In the case of Spotify, the data and analysis are automated—leveraging the massive amount of data generated to train and hone a prediction engine. Such algorithmic approaches allow for a diverse array of applications: from the search results at Google to product recommendations at Amazon to the advertisements placed on Facebook. Artificial intelligence and machine learning are central to the use of industrial robots, facial recognition engines, and autonomous vehicles. Autonomy, in particular, is growing by leaps and bounds. From farm equipment to industrial machinery to automobiles to drones, we are seeing increased application of algorithm-driven devices.

THE COMPLEMENTARITY BETWEEN PREDICTION AND JUDGMENT

Taken together, artificial intelligence, machine learning, and autonomy—coupled with such things as 3D printing and additive manufacturing—are driving what the World Economic Foundation has called the Fourth Industrial Revolution.[24] There is an increased blurring of lines between the physical and the digital. The classic battle of "hardware versus software" has been settled . . . and *software won*! Physical objects are being imbued with intelligence. Data and analytics are driving algorithms that are being used to improve operations, innovate new products and services, and deliver new value to stakeholders. In a 2021 McKinsey analysis, nearly a fifth of businesses surveyed reported refocusing their *entire* business around digital technologies. In general, the COVID crisis dramatically accelerated digital adoption.[25]

Given all of this, it may be tempting to assume that the age of smart machines is ascendant and that the age of smart *managers* is passing. I say not so fast. Agrawal, Gans, and Goldfarb, in their book *Prediction Machines*, highlight the opportunities *and* limits of artificial intelligence.[26] As we have discussed, artificial intelligence, and analytics more broadly, are ultimately about making better predictions—using the data at hand to develop a model that makes forecasts about future states of the world.

Artificial intelligence, however, is not capable of making better *judgments*—defined as the ability to make considered decisions or come to sensible conclusions. Judgment is a distinctly human endeavor. Ultimately, AI is only responsive to the objective function given to it by the coder. What is the correct objective function or a reasonable algorithm is a judgment of a human decision maker. What to do with any predictions generated and how they influence actual decisions is in the hands of the AI creator. Even with autonomy, the delegation of actions to a machine is, at the end of the day, a choice of a human agent.

A whole field of study has arisen looking at the ethics of algorithms. When should we or should we not allocate decision rights to machines? How do we interpret the "black box" algorithms that are generated with machine learning? What biases or prejudices may be amplified in a strictly data-driven approach to building models? In the world of autonomous vehicles, what rules or objective function should guide behavior in the case of a gut-wrenching choice to save either those in the vehicle or those walking on a city street? At the end of the day, all these questions require human judgment to answer.

Agrawal and colleagues point out that prediction and judgment are complements—each one enhances the value of the other. As with all complements, lowering the cost of one—say prediction (in essence what data and analytics *does*)—increases the demand for the other—judgment in this case. As a result, in a world with increasing efficacy in prediction, one would expect to see an increase in demand for judgment. Smart machines do not replace smart managers, they increase the demand for them.

My colleague Ed Hess has written eloquently about the types of "smart" managers we need in the digital age. In his book *Humility Is the New Smart*, he persuasively argues for a more humanistic approach to management.[27] The ubiquity of information available to us both on demand and in a moment's notice is devaluing classic "know-how." While being technology literate is a baseline need of all managers, the world is moving too quickly, with new technologies being developed and advanced almost daily, for expertise in the traditional sense to be a differentiator. What is far more valuable is how to learn, to adapt, and to make good judgments. Ultimately, managers need to think strategically as they respond to quickly evolving technology and hyperdynamic competitive markets. This is a theme we will return to in Chapter 5.

THE ETERNAL STRATEGIST'S CHALLENGE

The good news is that strategy in the digital age is amenable to analysis using many of the classic tools and approaches of strategy. In many ways, digital transformation is simply a new manifestation of the classic Strategist's Challenge (see Figure 1.3). The Strategist's Challenge is to envision and secure value-competitive positions at the intersection of the organization's values, the opportunities created in the marketplace, and the unique capabilities of the organization. An organization's values are its north star. They define the organization's overall mission and purpose, the domains in which it wishes to play (or not play), and the vision for its future. The market defines the opportunities available to the organization. What products and services do customers demand? What technologies are available? Who else is competing in each domain and in what ways? Last, an organization's capabilities determine how a firm may uniquely compete in the marketplace. What does the organization do well? What does it do better than anyone else? This is the key to understanding, identifying, and leveraging competitive advantage.

In simpler times, the Strategist's Challenge often centered around capturing scale and scarcity advantages. During the twentieth century, economies of scale were critical to driving down the costs of production in physical goods such as steel and automobiles. Owning scarce resources from land to raw materials to even technology, in the form of patents, was often the key to competitive advantage. Scale and scarcity still matter in the digital age, but the things that you need to scale and the scarcity you can leverage have largely shifted from the physical to the virtual—from hardware to software, if you will. Suddenly having the means for creating or capturing proprietary data has become the new scale and scarcity advantage.

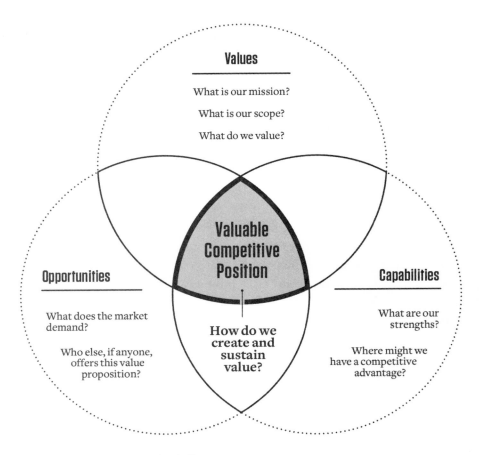

Figure 1.3. The Strategist's Challenge.

For many established companies, the digital age has created a particularly nasty version of the Strategist's Challenge. Suddenly, what made you successful no longer works. In other words, the opportunities provided by the market are moving away from your historic capabilities, threatening your established competitive position. Peter Drucker said it best: "The root cause of nearly every [business] crisis is not that things are done poorly. [It is that] the assumptions on which the organization has been built and is being run, no longer fit reality."[28] It is this misalignment between current capabilities and evolving market needs that leads to digital disruption. All too often, organizations fail to understand the radical nature of new digital technologies. Radical, in the technical sense, in that digitization renders existing technologies, business models, and capabilities ineffective. Prior success is no guarantee of future flourishing.

Thus, in the digital age, strategic thinking is absolutely critical. Fortunately, our strategist's toolkit provides several useful frameworks and approaches to begin analyzing a digitally transformed competitive environment.[29] An environmental analysis, sometimes called a PESTEL Analysis, helps answer the question, "What is happening in our competitive environment?" It is a framework for analyzing demographic, sociocultural, technological, political, legal, global, and macroeconomic trends. A competitor analysis is a foundational tool for identifying and collecting intelligence on your competitors. The key is to focus not only on your existing competitors, but your future competitors—to ask, "Who may leverage digital transformation to enter our market?" A capabilities analysis is a critical tool for analyzing what you do well and to identify which capabilities are likely to be a source of underlying competitive advantage. The challenge is to understand the people, processes, and systems that undergird specific organization capabilities and to analyze whether those capabilities are well aligned with one another and your external value proposition. Last, an analyst must ask whether that system of capabilities is both durable—maintainable and relevant looking forward—and inimitable—difficult for others to copy.

Together, these three tools—environmental analysis, competitor analysis, and capabilities analysis—provide a baseline to begin to strategize a digital transformation. The end of this chapter provides a framework to help you apply these tools to your organization. Consider Accor Hotels, a global provider of hoteling services. An environmental analysis would highlight important trends in the preferences of travelers, such as the desire for authenticity and unique experiences. Most critically, the analysis would highlight the trends towards home rental facilitated by online platforms. A competitor analysis would feature not only traditional competitors such as Marriott and Hilton, but intrepid entrants such as Airbnb—highlighting the different business models and capabilities inherent in each. Last, a capabilities analysis would help reveal what Accor is uniquely capable of achieving and begin to provide some clarity on how to position itself in a digitally transforming marketplace.

Of course, while these tools are helpful, they assuredly are not sufficient.

In this book, we augment our standard strategic analysis and toolkit with a deep dive into the underlying economic and competitive implications of digitization (see Figure 1.4 for our game plan). In the next three chapters, we attack the question, "What impact does digital transformation have on the nature of competition?" In Chapter 2, we focus on platforms—a ubiquitous feature of the digital world. We introduce the concepts of network externalities, enabling technologies, and winner-take-all markets. We discuss how organizations fight to establish dominant platforms and how the Competitive Life Cycle typically plays out as an industry evolves. In Chapter 3, we further unpack competing in the digital age. We highlight how platforms deconstruct the value chain, opening new opportunities for value creation and capture. We discuss various common competitive positions in digital markets and how to leverage the Competitive Life Cycle to innovate and build new sources of competitive advantage. In Chapter 4, we analyze appropriability—how to capture value from your efforts. We explore the new forms of value creation and business models catalyzed by digital transformation. We discuss the evolving role of intellectual property and the importance of complementary assets. We return to our Strategist's Challenge and explore how to best position your organization to create and capture value.

In the next three chapters, we discuss the managerial challenges and opportunities that the digital age poses. We tackle the question, "How do I digitally transform my organization to flourish in the digital age?" In Chapter 5, we focus on leading and managing in the digital age. We introduce the Digital Transformation Stack and discuss how to build out your digital infrastructure and how to recruit digital champions. Equally important, we explore how to build an agile organization, driving towards growth while remaining a humble leader. In Chapter 6, we deepen this discussion by focusing on the myriad of policy issues that arise in the digital age. We take a managerial perspective, discussing the strategic approaches firms may adopt to address the broad array of market and nonmarket forces having an impact on businesses in the digital age, ending with a discussion of the importance of values-driven leadership. Last, in Chapter 7, we discuss the transformative potential of digital technology, not only to create business opportunities, but to address pressing societal challenges. We discuss some emerging digital trends and, most important, provide a comprehensive roadmap for mastering your digital transformation.

Part I	What impact does digital transformation have on the nature of competition?

Chapter 2	Chapter 3	Chapter 4
Platforms	**Competing**	**Appropriating**
in the Digital Age	*in the Digital Age*	*in the Digital Age*
TOPICS:	TOPICS:	TOPICS:
network externalities, enabling technologies, winner-take-all markets, competitive life cycles	deconstructing the value chain, creating value, new business models, competitive positioning	intellectual property, complementary assets, innovation

Part II	How do I digitally transform my organization to flourish in the digital age?

Chapter 5	Chapter 6	Chapter 7
Leading	**Policy**	**Transforming**
in the Digital Age	*in the Digital Age*	*in the Digital Age*
TOPICS:	TOPICS:	TOPICS:
managing digital transformation, digital infrastructure, digital champions, agile organizations	perils of digitization, cybersecurity, data privacy, AI, antitrust, nonmarket strategies, values-driven leadership	promise of digital technology, climate change, smart cities

Figure 1.4. Structure of the Book.

At the end of each chapter, we present a framework specific to the concepts in the chapter with instructions on how to apply the framework to an organization—perhaps your own or one that you wish to analyze. These frameworks are designed to help you map out a digital strategy and to prioritize digital transformation efforts. They are meant to be integrative and cumulative, each providing specific insights that build and support analyses done in previous frameworks. In Chapter 7, we bring it all together, proposing a comprehensive set of steps, integrating across the frameworks, to develop and execute a digital strategy.

FRAMEWORK 1: SETTING A BASELINE

To begin your digital transformation, it is useful to establish a baseline map of the current competitive environment. We begin with an articulation of trends in the first column of Figure 1.5. This is in essence the environmental analysis (or PESTEL Analysis) referenced in the chapter. It is helpful to think of trends in broad categories: demographic trends, sociocultural trends, technology trends, political-legal trends, macroeconomic trends, and global trends. Pay particular attention to how the digital age may be having an impact on these various trends in the sector of the organization you are analyzing. For example, how are customers' preferences being shaped by emerging digital technologies, or how is the political-legal environment evolving to address emerging digital concerns?

Next, consider competitors in the second column. First, list the current primary competitors of the organization you are analyzing. How is each positioned in the market? What are their strengths? Do they have an identifiable competitive advantage? Pay particular attention to their digital capabilities. Have they begun their own digital transformation? Are they leveraging digital technologies to transform their product offerings or innovate new business models? Second, list potential, new competitors. This may include emerging new ventures or existing organizations looking to diversify into the target organization's space. For example, perhaps it seems reasonable, even likely, that a large "big tech" company will move into this space. Once again, for each of these potential competitors: What are their strengths? Do they have a competitive advantage? Perhaps they are digital natives—organizations formed with digital in mind. How can they leverage their capabilities to outcompete incumbents in this space?

Last, consider the capabilities of your target organization. What do they do particularly well? Dig deeper to understand why. How do they leverage people, processes, and systems to deliver those capabilities? For example, are their capabilities driven by superior talent secured through novel hiring processes? Or, perhaps robust IT systems allow them to be highly efficient in manufacturing. With capabilities unpacked, we need to ask three critical questions. First, are these capabilities well aligned, both internally and externally? In other words, do they mutually support one another to deliver on the value proposition of the organization? Second, are these capabilities difficult to imitate or are they inimitable? This is the

Trends	Competitors	Capabilities
Demographic	**Current**	**People**
What impact are demographic trends having on your sector, especially as it relates to digital technology?	Who are your current competitors? How are they digitally enabled? What is their competitive advantage, if any?	How do your people enable your capabilities?
Sociocultural		**Processes**
What impact are sociocultural trends having on your sector, especially as it relates to digital technology?		What processes enable your capabilities?
Technology		**Systems**
What impact are technology trends having on your sector, especially as it relates to digital technology?		What systems enable your capabilities?
Political-Legal	**Potential**	**Alignment**
What impact are political-legal trends having on your sector, especially as it relates to digital technology?	Who are your potential competitors in a digitally enabled world? How may they leverage existing strengths to compete in your sector? What is their competitive advantage, if any?	Are your capabilities well aligned with your value proposition to customers and partners?
Macroeconomic		**Inimitability**
What impact are macroeconomic trends having on your sector, especially as it relates to digital technology?		How hard would it be for others to imitate your unique capabilities that provide a competitive advantage?
Global		**Durability**
What impact are globalization trends having on your sector, especially as it relates to digital technology?		How valuable are these capabilities in a digitally enabled world?

Figure 1.5. Setting a Baseline Framework.

key to competitive advantage. Are there reasons to believe that these capabilities differentiate the organization from others in this space? Third, and
most important, do we imagine that any advantage these capabilities may
provide will prove durable? Often when we think of durability, we consider
whether the organization can maintain a specific capability moving forward. For example, a capability underwritten by having access to an exclusive talent, perhaps an elite athlete on a sports team, is only as durable as
that talent. It is also important, however, to think about whether a specific
capability will continue to provide value in the future. In the digital age, it
is often durability in this second sense that proves elusive. Will these capabilities continue to be valuable in a digitally transformed future?

PART I

WHAT IMPACT DOES DIGITAL TRANSFORMATION HAVE ON THE NATURE OF COMPETITION?

2

Platforms in the Digital Age

In 2009, the staid world of livery services—the paid transport of passengers by limos and taxis—was about to be upended. A new, San Francisco–based start-up, Uber Technologies Inc., was founded by Garrett Camp, a computer programmer, and Travis Kalanick, a serial entrepreneur. To say Uber came out of nowhere is somewhat of an exaggeration. In fact, dozens of companies identified the opportunity to leverage emerging mobile technology to redefine the livery experience. Two years earlier, Apple had released the iPhone, catalyzing the mobile revolution. Suddenly, everyone had a miniature computer in their pockets and with it the opportunity to redefine the paid passenger transportation experience.

The key insight was that mobile technology could be used to hail rides digitally. Originally conceived to connect existing limo drivers with passengers, Uber quickly recognized the opportunity to recruit new drivers. Cities that had a constrained supply of vehicles—in the case of taxis, often due to municipal regulations that created medallion systems to control the number of cars on the road—suddenly had an abundance of vehicles as new drivers signed up to Uber, Lyft, and other applications. This had the immediate effect of lowering the cost of livery services as supply increased, which created more demand for the service, incentivizing more drivers to enter the market, and ultimately driving a virtuous cycle greatly expanding the size of the market. In New York City alone, the number of trips provided by livery services grew from 479,000 per day in 2010 to more than 1 million per day by 2020; an increase driven by the introduction of ride-hailing apps.[1]

Fast forward nearly a decade: Uber reached a valuation of $82 billion in 2019.[2] Its annual revenues are greater than $10 billion, with over one hundred million passengers taking over five billion trips in over ten thousand cities around the globe.[3] The company has diversified into several adjacent businesses, including food delivery through UberEats, package delivery, and freight transportation. In each case, it pursued an asset-lite strategy, using mobile technology and software applications to connect customers with independent drivers, who bear the cost of purchasing and maintaining their vehicles. While Uber's rise has not been without controversy, there is no denying its transformational impact.

Uber is an example of a ubiquitous phenomenon—the platform. Uber didn't start a new limo service by buying cars, hiring drivers, and advertising on local radio. Rather it created a software application that connected people. Uber is a specific type of platform that we call a two-sided market marker—it connects numerous independent buyers to numerous independent sellers. Other examples include Airbnb, eBay, Amazon Marketplace, and Alibaba. In each case, the company serves as an intermediary to various partners in an exchange. This intermediary role allows them to scale quickly and, in many cases, lock out rivals who lack the scale to attract buyers and sellers.

Parker, Van Alstyne, and Choudary in their book *The Platform Revolution* define a platform as "a business based on enabling value-creating interactions between external producers and consumers."[4] They go on to observe that platforms create a participative infrastructure that "consummate[s] matches among users and facilitate[s] the exchange of goods [or] services." As they observe, the "platform is a simple-sounding yet transformative concept." We are surrounded by platforms in the digital age. An operating system (OS), such as Microsoft's Windows and Apple's MacOS in laptops and Apple's iOS and Google's Android in smartphones, is an obvious example. The OS serves as a platform on which other software applications are developed and deployed—connecting developers with users. Smart home voice assistants such as Amazon's Alexa and Facebook's Portal act like platforms connecting devices and allowing for software add-ons. Facebook, in general, is a prime example of a platform connecting users in its social networks and, more important from a revenue perspective, connecting advertisers with potential customers. Similarly, Google's ecosystem of search technologies helps connects advertisers with individuals and individuals with each other.

While platforms have most certainly proliferated in the digital age, the concept goes back centuries. Railroads could be thought of as platforms connecting buyers and sellers of goods across vast spaces. Electrical utilities serve as platforms for a vast array of electronic goods. The original telecommunications networks such as the telegraph and telephone share many traits with modern social networks. Network and cable television have long been platforms to connect content with viewers. Recent moves towards streaming services are simply the next iteration of these media platforms. Even the Internet itself can be thought of as a platform—a particularly valuable type called a general purpose technology that we will discuss later in this chapter—that has enabled many of the other platforms that have risen in the digital age.

THE CRITICALITY OF NETWORK EXTERNALITIES

Platforms are so important, and often disruptive, to existing businesses due to the frequent presence of *network externalities*. I always tell my students, if you remember only one thing from my course on strategy in the digital age, remember this concept. (Of course, I hope they remember far more, but you just want to make sure.) A network externality is defined as "a change in the benefit, or surplus, that an agent derives from a good when the number of other agents consuming the same kind of good changes."[5] Stated another way, a network externality is present when the value of a good or service increases with the adoption of that good or service by others. Consider Facebook. The value of its social network increases as more people use the service, increasing the likelihood of making connections with family and friends, or simply finding that long-lost high-school crush.

I think it is fair to say that at the heart of the platform revolution are network externalities. My favorite example is the telephone. What is the value of being the first person to own a telephone? Not much. You would have no one to call! But as people adopt the telephone, its value increases as you have more people to call. Similar dynamics are observed on other platforms. The value of using Windows OS increases as more people adopt

Windows-based personal computers. Each new adopter increases the likelihood that software developers will develop applications for the OS. Software is a classic high-fixed-cost, low- (or near zero) variable-cost business. Most of the expense is in developing the software before it is ever deployed. To cover their fixed costs of development, software developers want a large installed base to which to sell their software. Thus they are hesitant to customize applications for operating systems with low adoption.

The key to these network externalities is what economists Hal Varian and Carl Shapiro refer to as demand economies of scale—in essence, the larger a network, the more value created for the network's users.[6] Robert Metcalf, founder of 3Com, asserted what many now refer to as Metcalf's Law: the number of connections in a network grows nonlinearly as the number of nodes (that is, users) increases. Specifically, he argued that the value of a network is proportional to the square of the number of connected users on the network. For example, a doubling of users in a network from 10 to 20, leads to a four-fold increase in value (10^2 to 20^2). While some question the accuracy of Metcalf's Law and its applicability as networks grow extremely large, the general point holds that larger networks are preferable to smaller networks.

Network externalities can be driven by several forces. For example, in a social network like Facebook, we observe what is referred to as a direct or same-side user effect. Users gain value from the direct participation of other users. Waze, Google's traffic application, relies on user-provided updates on road conditions. This creates a same-side user effect: as more people participate, the more accurate become the predictions on road conditions. There are also same-side producer effects. For example, BitTorrent allows for peer-to-peer file sharing that enables users to distribute data and files over the Internet and whose capacity increases with the number of users. In fact, many industry standards, such as ethernet protocols or operating systems, demonstrate similar same-side producer effects.

Potentially even more powerful are indirect or cross-side effects. These are situations in which multiple producers connect with multiple users. Platforms that make such connections are often referred to as two-sided market makers as they connect both the producer and user sides of the market. Many of the most powerful platforms in the digital age capture such cross-side effects, from Google to Amazon to Uber to Apple to eBay and hundreds of others. As we've discussed earlier, network externalities can create a virtuous cycle in which an increasing number of users increases participation by producers, enhancing value-added services and/or lowering prices, catalyzing more demand by users. Airbnb is more valuable the greater the stock of accommodations available and the greater the number of potential renters seeking accommodations. The former drives more renters to the platform. The latter drives more people looking to rent their home. Together, they create increasing returns to scale.

As data has risen in importance, driven by the application of artificial intelligence and machine learning and the creation of digital algorithms, both same-side and cross-side network effects are growing. The very network externality that drives the scaling of a platform such as Airbnb generates vast amounts of data—everything from temporal changes in demand for specific geographies to the types of features that attract renters. This data can be aggregated to create value-added service for both producers and consumers. For those renting out their homes, this data can be critical to determining when to rent their home, at what price, and with what amenities. For renters, this data can help provide recommendations about where and when to schedule that dream trip.

Of course, it is important to recognize that not all network effects are positive. *Negative* network effects are possible and could lead to the quick collapse of a platform. Consider relationship matching services like Match.com or Tinder. If the platform is inundated by those deemed undesirable, then those that are attractive, however users define attractiveness, may leave the platform as they are inundated by requests from undesirable suitors. This, in turn, may cause moderately desirable mates to experience a similar dynamic and to exit the platform. Soon the service collapses as everyone exits. The same dynamic that led to Facebook becoming a dominant platform could lead to its unraveling if people suddenly decided that Facebook wasn't cool or, to update my vernacular, should be canceled due to its shortcomings. If you have teenagers, you may know that such a future for Facebook is not far-fetched. Fortunately for Facebook, it owns popular alternative platforms such as Instagram and WhatsApp.

THE IMPORTANCE OF ENABLING TECHNOLOGIES

Platforms, and the network externalities that drive them, often serve as enabling technologies. Enabling technologies are technologies that enable the rapid development of derivative technologies and solutions. The steam engine is a classic enabling technology that helped drive innovations across agriculture (tractors), manufacturing (power looms), and transportation (locomotives, steamships) and was a catalyst for the Industrial Revolution. The electric generator was a catalyst for the electrification of numerous machines, including the incandescent light bulb, at the turn of the twentieth century. The light bulb itself was an enabling technology transforming the way businesses and society functioned.

The most fundamental of enabling technologies are referred to as general purpose technologies. These are those rare technologies that have an outsized impact on innovation more broadly and often lead to massive increases in economic productivity. Richard Lipsey and Kenneth Carlaw propose that there have been only twenty-four general purpose technologies throughout history, including the wheel and the use of iron (see Figure 2.1).[7] In the digital realm, the computer and the Internet are typically cited as general purpose technologies. Clearly, they form the backbone of our modern digital world and have enabled numerous derivative innovations. While such general purpose technologies are rare, they can have enormous impact for generations.

Consider our example of Uber. Uber's technology and business model was only possible due to advances in mobile phones, especially Apple's release of the iPhone, which itself was driven by advances in computing and Internet connectivity. With regards to its UberEats service, the availability of inexpensive delivery has inspired several innovations in how and where foods are prepared and sold. For example, "ghost kitchens" have arisen that no longer have public dining areas and only serve customers via delivery. Who would have imagined that the whole business model of restaurateurs could be transformed due to digital technology when we were using our flip cell phones in the mid aughts!

In general, platforms can create numerous opportunities for downstream innovation. EBay's two-sided market platform began primarily as a way for individuals to sell unwanted goods around their homes—a virtual yard sale if you will. Quickly it became evident that there was a need for facilitating trustworthy monetary transactions online. Enter PayPal with its pioneering online payment system. For individuals wanting a secure way to ensure that payments were finalized between buyer and seller, PayPal filled an important need. A platform in its own right, PayPal would

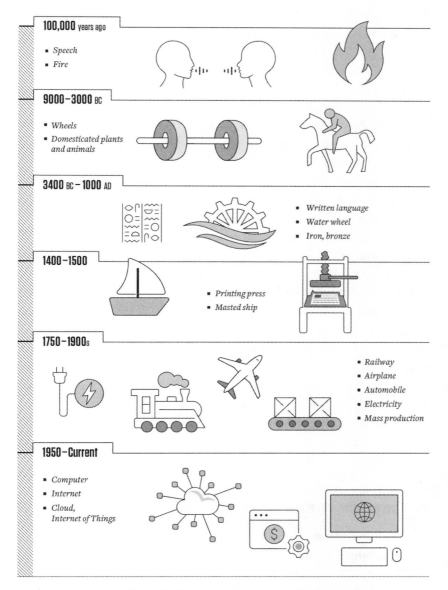

Figure 2.1. General Purpose Technologies. Source: Adapted from Spectrum Futures, https://spectrumfutures.org/will-5g-be-a-general-purpose-technology/.

eventually be acquired by eBay for $1.5 billion before being spun out as independent company, once again, in 2015.

This dynamic of a platform enabling the innovation of other applications and technology, some of them also platforms, is a common feature of our digital age. Consider, once again, operating systems. Microsoft's Windows and Apple's MacOS in personal computing and Apple's iOS and Google's Android in mobile computing have created vast ecosystems that have allowed thousands of software application developers to innovate new products and services. Many of these downstream innovators have created platforms of their own, such as Uber and Airbnb. To a lesser extent, gaming platforms such as Microsoft's Xbox and Sony's PlayStation have created a similar dynamic and encouraged the entry and innovation of numerous software titles while also facilitating business model innovation as gaming moved to an SaaS (software as a service) model.

Even traditional purveyors of physical goods are leveraging platforms to enable new innovative services and business models. For example, auto insurer State Farm has created an app that when installed on your mobile device tracks your driving behavior in your car and adjusts premiums and rewards safe driving on the basis of the data received. Consumer product good producers from Nike to Coca-Cola are leveraging digital applications to better engage with their customers, build brand equity, capture user data, and provide enhanced services and rewards. Amazon Marketplace provides third-party manufacturers and retailers the opportunity not only to broaden their sales bases but to innovate their product offerings in response to user data and feedback. Platforms tend to be multipliers for innovation.

THE RISE OF WINNER-TAKE-ALL MARKETS

The role of platforms as enabling technologies raises an important issue of who appropriates the gain from innovating new products and services. In the digital age, one of the opportunities—and, simultaneously, growing fears—is that platforms, and the network externalities powering them, are driving winner-take-all markets. A winner-take-all market is defined as one in which a single company comes to dominate the gains from trade in the market. In the extreme form, we may see a monopoly when a company captures 100 percent of revenues and drives all competitors from the marketplace. More likely are oligopolies, when a single firm may possess a dominant market share and generate outsized earnings compared to rivals who are left filling niche segments of the market.

We have seen the rise of a handful of "big tech" companies that command huge market shares globally or, at the very least, in large regional markets. Google commands over 90 percent of all global Internet search activity.[8] Microsoft's Bing is a distant second with less than 3 percent of all searches.[9] The one region of the world where Google does not dominate is China, where Baidu controls 80 percent of the market.[10] In online retail in the United States, Amazon captures close to 40 percent of all sales in an otherwise highly fragmented industry (second place Walmart has 5 percent market share online).[11] In China, Alibaba's Taobao and Tmall dominate e-commerce with over 50 percent market share by sales.[12]

In personal computing, Microsoft Windows continues to dominate with close to 80 percent global market share in desktop operating systems.[13] Surprising to some, Apple's MacOS remains a distant second with just 16 percent market share.[14] Similarly, Google's Android OS commands 72 percent of mobile operating system market share, with Apple's iOS collecting most of the rest with 26 percent.[15] While not quite a winner-take-all market, cloud infrastructure services is dominated by Amazon AWS (33 percent) and Microsoft Azure (20 percent).[16] Ride-hailing is dominated by Uber (37 percent) and China's Didi Chuxing (32 percent).[17] And so on and so on.

Their dominant positions allow these companies not only to extract much of the value created in their core markets, but also often to capture significant value from downstream innovators leveraging their platform. For example, Apple charges a 30 percent commission on the total price paid for apps and in-app purchases on its phones—something that has become a significant point of contention with software developers such as Epic Games, maker of the popular multiplayer game *Fortnite*. Millions of third-party retailers pay fees to Amazon to use its Marketplace platform—generating roughly 50 percent of all Amazon e-commerce sales.[18]

Perhaps most disconcerting, we are seeing these big tech companies horizontally diversify, moving into adjacent markets by leveraging their platform dominance and digital expertise. Apple and Facebook have moved into the payments space with their Apple Pay and Facebook Calibra digital wallets, respectively. Google became a major player in home HVAC and security systems with its purchase of Nest in 2014. Amazon made a significant move into physical retail stores with its purchase of Whole Foods in 2017. Google, Amazon, Uber, and Apple are all investing in autonomous vehicles, posing a threat to existing automobile manufacturers.

There are several factors driving these winner-take-all dynamics. In some cases, such as cloud infrastructure, traditional economies of scale largely explains the dominance of a handful of players. Building a successful cloud infrastructure business requires significant capital investment to build data centers. Google, which manufacturers its own servers for its data centers, claims to be one of the largest computer makers in the world. Software, in general, is a classic high-fixed-cost, low-variable-cost business that benefits from spreading the fixed cost of development over large sales bases. Scale matters in software and conveys significant advantages to those who can grab large market share.

More often, the very network externalities that drive platforms, drive the tendency towards a single dominant player. If everyone is signed into and using Facebook, it is very hard for a competitor to gain market share when everyone is already on Facebook enjoying the benefits of its large network. Drivers and riders know that Uber has a vast network of users, increasing the likelihood of finding a car or a rider. Nascent ride-hailing services simply can't compete with the advantages of an entrenched rival like Uber. As a result, we often see extreme shakeouts, when players are driven from the market as a rising platform company catalyzes the virtuous cycle of network externalities to crush the competition. This creates challenges for antitrust regulators, as the tendency towards monopoly, or at least quasi-monopoly, is driven by the underlying dynamics of digitization and may be hard to eliminate entirely. (This is something we will revisit in detail in Chapter 6.)

Compounding the impact of network externalities are further barriers to competition that may be generated as a firm and its platform scales. For example, complementarities between hardware and software or between applications can raise switching costs for users and drive lock-in to a specific platform. Anyone who is part of the broader Apple ecosystem understands this well. Owning an iPhone encourages the use of iCloud, which creates a lock-in effect as the cost of switching all your photos, videos, and applications to a new platform (say Android) discourages such changes.

Layer in the use of a Mac and an Apple Watch and suddenly your digital life is dominated by Apple.

The overall tendency towards winner-take-all markets has become increasingly the case in our data-driven economy as, discussed earlier, we move from connectivity to aggregation. Data creates a virtuous cycle in which more data leads to improved algorithms, lowering transactions costs and/or improving value add for users, driving more transactions, generating yet more data to further improve algorithms (see Figure 2.2 for an illustration). This process drives scale, leading to dominant companies. Companies such as Amazon, Google, Apple, and Facebook possess mountains of data on your individual interests, preferences, and behaviors. This data can be used to create reasonable predictions on your future actions, leading to products and services engineered to best meet your individual needs, while continuing to generate new data to predict your behavior even better.

In such a world, the constant threat of disruption might not be enough to unseat currently dominant companies and platforms. Innovating a new technology or application may not be enough to overcome the inherent advantages that data provides to established rivals. Say you come up with a superior ranking approach for Internet search results. Google's vast investments in servers and its collection of data mapping out the linkages between web pages may simply be too big an advantage for an innovative upstart to overcome—especially when the dominant player has the resources to simply copy you, acquire you, or, even worse, kill your business using aggressive tactics (a potential antitrust concern that we will discuss in Chapter 6).

THE QUEST FOR PLATFORM DOMINANCE

If to the winner go the spoils, a logical question is how does one create a dominant platform, especially in markets not already dominated by an entrenched competitor? For every Facebook, there is a MySpace that tried valiantly but ultimately failed to create a dominant platform. Actually, there are usually a few hundred MySpaces that try and fail for every successful Facebook. We often suffer from success bias, recalling the winners and underestimating the challenges to establishing a dominant platform. Facebook was certainly not the first social media site: MySpace was

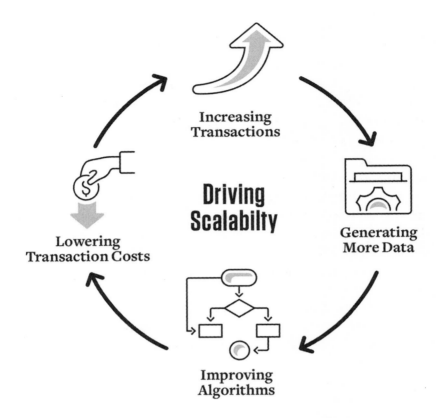

Figure 2.2. The Virtuous Cycle of Data. Source: Adapted from Ron Adner, Phanish Puranam, and Feng Zhu, "What Is Different About Digital Strategy? From Quantitative to Qualitative Change," *Strategy Science* 4, no. 4, December 2019, 253–261.

founded in 2003, a year before Facebook debuted. In fact, a largely forgotten site called Six Degrees debuted six years earlier in 1997. So why did Facebook win the quest to dominate social media? Was Facebook a better product? Better managed? Did it simply get lucky?

Critical to success is solving the platform "chicken and egg" problem. Users don't want to frequent applications that are not used by other users, so how do you attract those initial users? For two-sided market makers, the problem is especially pronounced: users will not flock to a platform if producers are not available, while producers are unwilling to participate if users are not available. Consider Airbnb. To attract renters, there needs to be a sufficient stock of homes for rent on the platform. To attract those willing to rent their home, there needs to be a sufficient pool of renters to make it worth their time and effort to post. Do you try to build a sufficient base of home providers first? Perhaps, you try to attract users first? Both, simultaneously?

One key is to lower the barriers to initial adoption. There are several strategies companies use to attract those critical first users and producers. Many platforms start off by giving free access to users, producers, or both. Some adopt "freemium" business models that only charge fees to those who upgrade to premium services. Others adopt advertising-driven business models, relying on third-party advertisers to monetize the platform, keeping it free for users. Of course, good old-fashioned marketing and advertising can help drive traffic to a new platform. Some companies rely on promotions and even payments to users to incentivize use of their platforms. For example, PayPal paid new customers $10 for signing up and $10 for referrals in their early days. When launching in Seattle, Uber paid limo drivers even when they were idle to help incentivize them to join the platform.

Of course, as important as attracting initial users is keeping those users on the platform. Parker, Van Alstyne, and Choudary identify three general strategies that firms utilize to establish a dominant platform that they refer to as pull, facilitate, and match.[19] A pull strategy refers to those strategies discussed above to attract, or pull, users to the platform. A facilitate strategy refers to ways to encourage interactions and use of the platform once users are signed up by providing tools and value-added services. Especially valuable are services that provide single-user utility, in other words, value even if no one else signs up to the platform. For example, Instagram provides a way to store your photos even if you don't share them. Last, a match strategy is often at the core of platforms—finding ways to connect users, or users and producers, in meaningful and valuable ways. Data and analytics are often at the core of these match strategies. This is how a platform moves from mere connectivity to aggregation.

One of the critical questions to answer is whether to make your platform more proprietary or more open? Proprietary platforms, sometimes referred to as closed systems, are typically incompatible with other platforms and standards. They are, in essence, "walled gardens" where participants agree to the terms and conditions set by the platform's creator.

Apple is a prime example. The MacOS and iOS are proprietary operating systems completely controlled by Apple. The only devices available with the MacOS or iOS are Apple products such as the Mac, iPhone, and iPad. While third parties may develop software and applications for Apple, they do so under the rules dictated by Apple, including selling only through Apple's App Store.

In contrast are more open systems that allow for, and even encourage, joint ownership and control. Google's Android is an open architecture. The core code was developed by Google through its Android Open-Source Project, under which people can take the code and create customized operating systems. While you can create applications and make them available through Google's app store, Google Play, you can also install apps from other sources. Unlike Microsoft, which sells its Windows OS to hardware companies such as Dell and HP, Google provides Android for free, choosing to monetize the OS through data and advertising.

Of course, open or closed is often not a binary decision. A continuum of degrees of openness or closedness may exist. When Apple debuted the iPhone, it adopted a very closed system that did not support third-party native applications. Downloading a non-approved app could lead to the phone becoming a "brick"—basically corrupting the operating system and making the iPhone a very expensive paper weight. The outcry from frustrated users and developers nearly killed the iPhone in its infancy. Apple relented a year later, creating a software development kit, helping facilitate third-party developers, and launching the App Store.

At the heart of the question of how open to make a platform is a series of trade-offs. On one hand, proprietary systems give you tight control of the user experience. Apple's closed architecture allows them to create a high-end differentiated product that is beloved by users and for which they can charge a substantial premium. Furthermore, a closed system often raises switching costs and encourages lock-in by users, creating barriers to competition. However, that level of control can come at the expense of easing platform adoption. Uber had to grapple with whether to allow drivers to register with rival platforms such as Lyft. Clearly, having drivers be exclusive to Uber would be a huge competitive advantage. However, removing optionality would discourage drivers from signing up with Uber in the first place.

Similar trade-offs exist with regard to who controls data created by the platform. The platform provider would likely prefer exclusive access to any data captured about users' preferences and behaviors. However, that data may be critical to other partners using the platform. This is especially true for two-sided platforms where producers may view such data as a critical reason to join the platform in the first place. Consider education upstart Coursera, the largest provider of massively open online courses (MOOCs). Its business model is to partner with accredited universities, some of the top ones in the world, to create online content that it then distributes through its platform. (In full disclosure, my home institution, the University of Virginia, was an early partner with Coursera and I was one of the first professors to offer a course on Coursera.) Its university partners would love to know more about not only who is subscribing to its courses, but what courses people are subscribing to in general. This type of competitive intelligence is highly valuable. Coursera could hold tightly to this information, maybe selling consulting services to its university partners. However, such behavior could discourage universities from using the platform in the first place. At the end of the day, Coursera chose a nuanced strategy, sharing some, but not all, of its data.

A classic illustration of the pros and cons of openness and control was the personal computer "PC wars" of the mid-1980s. On one side you had IBM, the venerable maker of computer hardware pushing a standard featuring a Windows OS and an Intel processor—what came to be known as the Wintel standard. On the other side you had Apple with its proprietary MacOS. IBM made the tactical decision to open up its architecture to what were referred to at the time as "clones," PCs made by the likes of companies such as Compaq, Gateway, HP, and Dell. While only a Mac came with the MacOS, these manufacturers were allowed to make Wintel-based machines. IBM's open strategy proved the winner. The massive supply of Wintel-based PCs helped the IBM-backed platform capture network externalities and powered Windows to a dominant position for desktop and laptop operating systems. Apple was relegated to a distance second place— as low as 3 percent market share by the late 1990s, a share so low that Apple was finding it hard to cover the development costs of its proprietary strategy (a new version of the MacOS can cost well above $1 billion). In fact, many pundits were forecasting the imminent demise of Apple, until Steve Jobs orchestrated one of the great corporate turnaround stories of all time with the introduction of the iPod and the iPhone.

In the end, while IBM's open strategy won the platform battle, a critical choice ultimately cost it. IBM believed its superior hardware design and manufacturing capabilities would serve as the basis for outcompeting rivals. While it may have tried to acquire a fledgling Microsoft (though a consent decree with the Justice Department due to antitrust concerns may have made this unlikely), it ultimately decided to simply partner with Microsoft. This proved fatal, as IBM eventually exited the market, selling its PC business to Lenovo as it pivoted away from what had become the low-margin business of assembling Wintel machines. Meanwhile, Microsoft would grow to be one of the largest companies in the world, with a market capitalization of over $2 trillion in 2021.[20]

The IBM-Apple saga highlights an old Silicon Valley trope that "software eats hardware for lunch." Since software is often the critical element of a platform and the corresponding network externalities, the control of software is often far more valuable than the production of hardware. IBM may have been right to open its platform, but it failed to control the most valuable piece—the operating system. This will likely become more of an issue as more devices are embedded with intelligence, from televisions to refrigerators to home security systems to cars.

In general, in the digital age, identifying who your partners are and how you would like to engage them is critical. It is interesting to note that increasingly your partners may also be your competitors, in either the past, the present, or the future. Netflix uses Amazon Web Services as its cloud infrastructure provider. It also competes directly with Amazon Prime Video. Google Maps and Google Search are ubiquitous on Apple iOS devices. Amazon's Alexa voice assistant can be used to control Google's Nest products. In some cases, companies establish consortiums with competitors to craft mutually beneficial ecosystems. For example, seven large European banks, including Deutsche Bank and Rabobank, formed the Digital Trade Chain in 2017 to build "a blockchain-based platform design to facilitate cross-border trade for small and medium-sized businesses."[21] We will return to partnership strategies in the next couple of chapters.

THE DYNAMICS OF COMPETITIVE LIFE CYCLES

So how do we begin to strategize around competing in a world dominated by platforms? The challenge of establishing a dominant platform highlights the importance of understanding the dynamics of competition. Fortunately, there are some common patterns to how technologies and industries evolve that define what we refer to as the Competitive Life Cycle. These patterns can help us to identify time-dependent strategies that match the stage of technological development and evolving market conditions with the specific capabilities that an organization may possess or could develop. In the most basic sense, we are adding competitive dynamics to our classic Strategist's Challenge. As market opportunities and technologies evolve, how do we identify and capture valuable positions that are consistent with our organizational vision and mission and leverage existing or developing capabilities?

Consider the canonical Competitive Life Cycle, illustrated in Figure 2.3. Early in the life cycle of a new technology, what we refer to as the Emergent Phase or sometimes referred to as the Era of Ferment, questions abound about the viability of the technology. (Note: we are using the term *technology* broadly to refer to what might be new products, services, platforms, applications, or even business models.) R&D may seem relatively unproductive in this phase, as various technological trajectories are experimented and abandoned. The technology will likely be inferior compared to existing products on the market—either functionally limited or too costly. To the extent the technology finds adoption, it is often in niche applications in which the technology creates outside value sometimes for underserved customer segments.

During this phase, it is often entrepreneurs who are pioneering and pushing the technology forward. There are several reasons for this. New ventures have the luxury to experiment free of the bounds of maintaining existing revenue streams and pleasing existing customers. They are free from concerns of cannibalizing existing products or services. Positive profits can be hard to come by, requiring a certain degree of patience. Their sheer potential numbers allow for rapid experimentation pursuing numerous technological trajectories. Most of these entrepreneurial efforts are destined to fail. In fact, over 90 percent of all ventures in the United States end up failing.[22] Once again, we tend to remember the Facebooks of the world, failing to recognize the hundreds of failed social media companies that entered the market around the same time. Venture capitalists know this well and typically look to hit "home runs"—those outsized successes that can compensate for the losses on the rest of their portfolio.

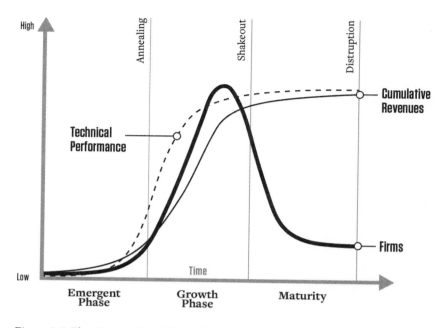

Figure 2.3. The Competitive Life Cycle.

As the market evolves, successful technologies typically enter an "annealing" phase transition. *Annealing* is an engineering term referring to the process by which metals are hardened into a set of desired properties. In our context, we use the term to refer to how technologies often evolve towards a dominant design—a set of common design features that come to dominate the market. At the outset of the auto industry at the turn of the twentieth century, there were numerous technology trajectories being pursued, including the gasoline-powered internal combustion engine, electric battery vehicles, kerosene-powered cars, and even the steam-powered Stanley Steamer. Eventually, the internal combustion engine came to dominate, providing the fundamental architecture of cars for over a hundred years. (Which only now is being disrupted by the return of battery-powered electric autos.)

In the digital age, the annealing phase transition usually involves the establishment of platforms and the emergence of network externalities. Apple was a relatively late entrant into the smartphone market, entering years after pioneers such as Palm and Blackberry. However, the innovative design of Apple's iPhone came to define the market and establish the dominant architecture for mobile devices. Once established, Apple began reaping the benefits of network externalities that increased switching costs and locked in customers to their platform. Even the internal combustion engine could be viewed as a platform that captured network externalities as gas stations began to proliferate to meet the increasing demand for gasoline.

As a dominant design is established, we typically enter the Growth Phase of the market. The Growth Phase is marked by an exponential increase in adoption—the sweet spot of the adoption S-curve illustrated in Figure 2.2. Note that the technology itself usually progresses along an S-curve as well. Suddenly, what may have seemed like a nonviable technology, or at best a niche technology, is challenging existing market offerings. Perhaps costs have decreased due to economies of scale or learning curves. For a lot of digital technologies, it may be that user and producer values have increased as more people adopt a platform.

As revenues increase during the Growth Phase, new entrants are attracted, including not only entrepreneurs but incumbent firms within the industry and firms in adjacent industries that diversify into the segment to try to capture the emerging market opportunity. This is typically when you start hearing about the disruptive potential of the technology. Suddenly, existing market players fully awaken to the transformative impact of the new technology. The new technology may be disruptive in the technological sense, rendering existing technological know-how useless—what is often referred to as radical innovation. Alternatively, the new technology may be an architectural innovation—not necessarily a major technological change, but one that challenges existing business models and renders complementary capabilities useless. Many platform innovations pose this latter type of disruption. Consider how Uber and Airbnb have transformed livery and hoteling services, respectively, without fundamentally changing the core technologies for either industry.

During the Growth Phase competition intensifies. Profits may continue to prove elusive as firms fight to establish dominance—investing in further innovation, offering value-added features to differentiate, slashing prices, or even paying users to gain market share. This is especially pronounced in winner-take-all markets. Facebook was probably well served to focus on user acquisition during the growth phase of social media and to worry less about monetization and profitability. The key was to capture the network externalities that resulted from growing its network of users. It helps explain why a company like Amazon went decades without posting consistent profits as it built out its business, and why a company like Uber was still struggling to turn a profit as of 2021.

The rising importance of data in the digital age only intensifies these dynamics. Increasingly, success rests on capturing a robust and powerful set of data that allows you to refine offerings and create more customer value (what we referred to as "aggregation" in Chapter 1). Creating virtuous cycles by attracting users to capture data that you then use to refine product features and offerings, perhaps using machine learning, is becoming the key competitive advantage. This is true for virtually all the big tech companies such as Google, Apple, Amazon, Facebook, and Microsoft. It is true for more specialized digital players such as Uber, Airbnb, and Netflix. It is even true for established players in retail (Walmart), automotive (GM Cruise), banking (Citibank), and so on.

During the Growth Phase, we often see the emergence of hypercompetition, defined by Dartmouth professor Richard D'Aveni as rapid and dynamic competition characterized by unsustainable advantage.[23] The fortunes of individual companies may ebb and wane quickly as the battle for dominance unfolds. The result is often a shakeout—when many competitors exit the market either through bankruptcy, forfeiture (in the case of diversified firms), or acquisition by others. The prevalence of shakeouts is profound. Almost every successful tech company today has left in its wake hundreds of unsuccessful competitors. Apple's iPhone and Samsung adopting Google's Android OS have come to dominate smartphones, leaving behind the likes of Nokia, Microsoft (RIP Windows Phone), Blackberry, and Palm. Before Google, there was Yahoo!, AskJeeves, Lycos, Altavista, and countless others. These lists go on seemingly forever.

The Growth Phase is when critical decisions need to be made about whether to design platforms to be more open or closed, whether to partner and with whom, and how best to capture value. I am sure many managers would love a simple answer to these questions. Strategy, however, is not about identifying best practices, if such a recipe could even exist. Strategy is about identifying unique positions that leverage your capabilities and that you can defend from rivals. For some firms, the main strategic question is simply how to become the dominant platform. For many others, the best answer may be to recognize that platform dominance is unlikely and to look to other parts of the value chain where they can leverage unique capabilities to capture value.

In Chapters 3 and 4, we will unpack these questions. Chapter 3 will focus on some of the key questions one needs to grapple with while competing in the digital age, including how do you think of your partners and where do you position yourself in the value chain? Chapter 4 will raise again the critical issue of appropriability—how do you create and capture value from innovation? For now, the critical thing is to identify where are you in the Competitive Life Cycle. Are you in the Emergent Phase, when rapid experimentation and innovation are key? Are you in the Growth Phase, when competition intensifies and the fight for platform dominance is prominent? Are you in the Mature Phase, when growth slows, and competition often shifts to production efficiency and service? Is your sector poised for a new disruption, starting the Competitive Life Cycle all over again?

FRAMEWORK 2: ANALYZING THE DIGITAL ENVIRONMENT

With our baseline snapshot of the current competitive environment established using our first framework, we can now pivot to the dynamics of competition and understanding where we are in the Competitive Life Cycle (see Figure 2.4). In each column, check the box that best captures the current state of the industry and organization you are analyzing and provide a description of the current state of affairs in that space (similar to the sample text provided here). This will be useful as we consider strategic positioning for the organization in later frameworks.

We begin with the industry stage in the first column. Has a new technology or business model emerged within the sector being analyzed? If so, how developed is the new technology or business model? If adoption is limited and there is still significant uncertainty about how products and services may eventually look, then you are likely still in the Emerging Phase. If adoption is accelerating and there is increased certainty in what the dominant design may look like, you could be entering the Growth Phase characterized by significant entry by competitors. If a shakeout has begun, with some new entrants and established incumbents struggling and exiting the market, you are likely entering a Mature Phase marked by

Industry Stage	Organization Stage	Industry Archetype
☐ **Emerging**	☐ **New Venture**	☐ **Dominant Platform**
A new technology epoch has begun. The new technology or business models have only a small foothold in the market. There remains much uncertainty about how the new technology may play out.	A relatively new organization. May be privately held and/or venture financed. Typically, more nimble than other competitors but lacking in needed complementary assets.	Industry in which a dominant platform already exists or there is a potential platform play. Marked by significant network externalities and the potential for winner-take-all dynamics.
☐ **Growth**	☐ **Growth Enterprise**	☐ **Segmentable**
New technology or business models are accelerating in adoption. A dominant design is beginning to emerge. A competitive shakeout may have begun or is imminent.	An organization experiencing significant top-line growth. May be experiencing some growing pains from scaling. Potentially establishing a sustainable competitive advantage.	An industry in which there may exist multiple platform plays catering to distinct market needs.
☐ **Mature**	☐ **Mature Competitor**	☐ **Protected**
The growth rate of adoption of the new technology or business model is beginning to slow. Competition is becoming more stable, but more intense.	A well-established competitor that is experiencing relative stability in its revenues and cash flows. May have significant capabilities that provide competitive advantage in the current market.	An industry in which there exist potential barriers to competition even in the presence of platforms, such as strong intellectual property protection like patents, geographic segmentation, or economies of scale.
☐ **Decline/Disrupted**	☐ **Troubled**	☐ **Competitive**
The current technology epoch may be reaching the end of its life cycle. New technologies and business models are beginning to emerge. Established businesses may find revenues and profits under pressure.	A competitor that is struggling to maintain revenues and cash flows. Perhaps once a growth enterprise and/or mature competitor that is finding it increasingly challenging to compete.	An industry marked by weak barriers to competition. Entry by competitors is relatively easy. Competitive success often centers around leveraging complementary capabilities.

Figure 2.4. Analyzing the Digital Environment Framework.

a decrease in the rate of growth in the new dominant design. Last, if the market is relatively stable with little turnover in competition and limited growth in revenues, you may be entering a phase ripe for disruption by new technologies and business models—perhaps some potential candidates for disruption have already started to appear.

In column two, organization stage, we address the organization's life cycle. This should be relatively straightforward. Is the organization relatively new? For entrepreneurial ventures, are they still private held, perhaps backed by venture capital? In the box, provide details on the organization's current maturity. Is the organization experiencing significant top-line growth? Then it may be more of a growth enterprise. What pain points are they experiencing as they grow? Is the organization beginning to establish a competitive advantage in the marketplace? Alternatively, has the organization been around for a while? Perhaps it is a mature competitor with relatively stable cash flows and well-established capabilities. Or it may be a troubled organization facing increasing competitive pressures and declining revenue and profitability. In all cases, use the box to describe the organization's current competitive situation.

Last, in column three, reflect on the evolving industry archetype. Recognize that any mapping onto an archetype may be both speculative and incomplete. Such an exercise is about stimulating your thinking on what competition may look like in the future. A dominant platform archetype is a sector already dominated by a dominant platform or that may be subject to a platform play in the future, likely characterized by network externalities and potentially a winner-take-all dynamic. Segmentable markets may be the most common archetype, in which different organizations position themselves to cater to different customer needs. In the digital age, this may include markets in which there are multiple platforms in play and multiple platforms will likely persist in the future. Protected markets are those that afford some ability to erect barriers to competition, even in the presence of platform plays. This may include various forms of intellectual property protection, such as patents, or other forms of competitive advantage, like scale or location advantages. Last, competitive markets are those characterized with limited barriers to competition and easy entry by competitors.

In all cases, use the boxes to summarize some thoughts on what competition is likely to look like in the sector. In the next chapter, we will use this analysis of the digital environment to help inform organization-specific strategies. You can imagine, for example, that the best strategy for a new venture facing a growing market in which a dominant platform is in play may be completely different from a mature organization facing an emerging market that allows for segmentable positions.

3

Competing in the Digital Age

In October 2020, people in Phoenix, Arizona, were greeted with a truly futuristic sight—fully driverless vehicles plying around the streets, ferrying passengers to their destinations. Waymo One was the latest offering from Waymo, the wholly owned autonomous vehicle (AV) subsidiary of Google parent Alphabet. Waymo had begun as the Google Self-Driving Car Project back in 2009. The brainchild of Silicon Valley inventor and entrepreneur Sebastian Thrun, the project was originally housed in Google's secretive X-Labs. The vision of a self-driving car had seemed a quixotic pursuit, yet in little over a decade incredible progress had been made. By January 2020, Waymo vehicles had driven over twenty million miles on public roads and tens of *billions* on simulated roads.[1]

Competitors had taken notice and were frantically developing their own autonomous vehicles. General Motors went out and acquired Cruise, a leading start-up in the autonomous space. Ford hired away several former Google X employees to spearhead their AV efforts. Audi formed partnerships with Aurora, an AV software start-up, and Nvidia, a Silicon Valley chipmaker. Electric vehicle pioneer Tesla was pushing its own vision for self-driving cars, including autopilot features in its vehicles, and collecting data on the millions of miles driven by its customers. Volvo, Nissan, and Jaguar initially invested in AV projects of their own, only to eventually partner with Waymo.

Entry was not limited to existing auto companies. Apple had long been rumored to be working on its own autonomous vehicle.[2] Uber had hired away four faculty and thirty-five technical staff from Carnegie Mellon University's National Robotics Engineering Center to launch their own efforts.[3] Amazon had long been experimenting with autonomous delivery vehicles only to go out and acquire Zoox in 2020 and enter the on-demand autonomous ride-hailing space, competing directly with Waymo. Even venerable chipmaker Intel was getting into the AV race, going out and acquiring Israel start-up Mobileye.

The rise of autonomous vehicles was obviously disruptive and threatened to upend the existing competitive market for automobiles. This was not only a new product or service but a technology that promised to fundamentally alter the nature of the competitive game in automobiles. Suddenly, a traditional manufacturing business in which firms competed for brand differentiation and economies of scale was being supplanted by a digital world characterized by platforms and network externalities. Data was becoming critical, as the heart of autonomous vehicles was using machine learning to generate algorithms based on trillions of data points. To the Silicon Valley intelligentsia, software was going to eat hardware for lunch.

How should one compete in such a digitally transformed world? As a new venture? As a diversifying tech company? As an incumbent automaker? Where should you position yourself in the value chain? Do you go for creating a dominant platform or do you pioneer other positions perhaps more aligned with your capabilities? Ultimately, how can you best create and capture value? What business models are available for monetizing your offerings? How may you secure a competitive advantage over rivals? This chapter is about understanding and unpacking the evolving competitive game in the digital age and designing effective strategies given your unique position and capabilities.

THE DECONSTRUCTION OF THE VALUE CHAIN

As discussed in the last chapter, the digital age has been marked by the rise of platforms, the ubiquity of network externalities, and the frequent presence of winner-take-all markets. This can make it seem as if competition in the digital age is a Darwinian battle to be the last platform standing in an otherwise barren competitive landscape. However, the impacts of digital technology can be subtler and, in some ways, even more insidious than that. Digitization can lead to the deconstruction of traditional value chains. In other words, the basic business architecture of how various economic players are organized to deliver value to end users can be upended. Suddenly, a new set of suppliers and players in adjacent industries may be delivering important pieces of your value stack.

The value chain refers to the full range of activities, the interlocking exchange of goods and services, needed to deliver a product or service. The business architecture refers to the set of players—such as suppliers, downstream manufacturers, and complement providers such as software for hardware or services for products—that interact together to ultimately deliver a product or service. On one extreme, we have fully vertically integrated businesses in which the whole suite of activities in the value chain occurs within the boundaries of one firm. On the other extreme, we have virtual organizations that outsource almost all activity in the value chain, potentially including manufacturing, design, and R&D.

Consider once again the auto industry (see Figure 3.1). Traditionally, auto manufacturers have developed relatively proprietary business architectures marked by owned assembly, vertical integration into engine manufacturing, outsourcing of most other parts to dedicated suppliers, and formal exclusive partnerships with downstream dealerships which serve as both retail channels and service providers. There are several reasons for such an architecture. First and foremost, integration may help reduce transaction costs. Transaction costs refer to those costs incurred to secure an economic transaction beyond the direct costs of purchasing a good or service. These may include negotiating contracts or verifying the quality of a good or service received. For example, for an auto company, choosing to tightly partner with a dealership can help you better monitor whether it is providing high-quality maintenance on your vehicles and reduce the risk that it will try to take advantage of your brand and exploit customers by giving them shoddy service.

Underlying most transaction costs are information asymmetries between buyers and sellers that raise the risk of opportunistic behavior by one party of exchange and results in costly efforts to minimize those risks. By vertically integrating, or at least forming tight partnerships, companies can help reduce these risk-mitigation costs. It is interesting to note that, in general, digitization has helped lower transaction costs globally. Now

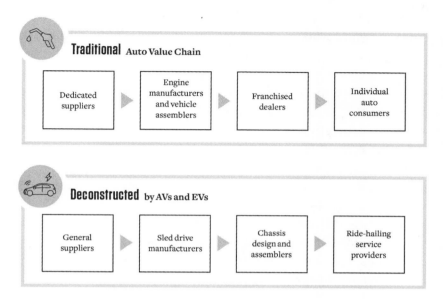

Figure 3.1. Deconstruction of Auto Value Chain.

you can monitor in real time suppliers that are halfway around the world, perhaps using sensors and cameras to observe production and leveraging artificial intelligence to process the vast amounts of data generated to identify production concerns. Increases in bandwidth, storage, and processing power and the subsequent ability to monitor partners that comes with them have arguably been central to the increased prevalence of outsourcing globally and have led to the rise of nearly virtual firms in a wide variety of industries.

So, let's consider what impact autonomy may have on the auto industry. Downstream, we may see the rise of direct-to-customer distribution channels bypassing the need for extensive dealer networks. In the same way that Dell Computer bypassed physical retailers by pioneering online made-to-order PC sales, auto companies could leverage online interfaces to design, order, and deliver custom vehicles. In fact, we are already seeing Tesla push this model in electric vehicles, bypassing the need for building out an extensive retail network and reducing the need for inventory.

On the supply side, we may see vertical disintegration between key components like the engine and chassis. In the case of electric vehicles, designers have discussed the possibility of "sled" designs for which the entire drive train—batteries and electric motors—are produced in a standard eighteen-inch base allowing for an infinite number of chassis configurations to sit on top of the base. This may very well lead to specialization, when certain companies will focus on producing the base—perhaps optimizing battery life and range—while others focus on custom chassis design catering to individual customers' specific tastes. Such decoupling will likely be accelerated by autonomy as software becomes the "brain" of the car and comes to define the riding experience. Soon you may be purchasing a custom-designed vehicle from an Italian design shop that leverages a mass-produced drive train manufactured in South Korea and autonomous software from Waymo.

Of course, the more transformational possibility of AVs is that the whole consumption model of automobiles is upended. Suddenly, individuals will not need to own their vehicle. Rather they will simply purchase ride-hailing services on an as-needed basis. Car utilization will dramatically increase as each individual automobile picks up and drops off multitudes of passengers a day. This ultimately will significantly reduce the number of cars needed on the road, not to mention dramatically reduce demand for things such as parking. In such a world, competition could look very different. Rather than millions of customers looking to purchase vehicles, we could have a handful of companies looking to be AV ride-service providers—companies whose business model is not to sell cars but to sell a service, perhaps charging a subscription fee or using advertising during the ride to monetize the service. Given the network externalities in play as discussed in Chapter 2, this could start to resemble a winner-take-all market in which one or two companies come to dominate globally.

As the autonomous vehicle case illustrates, the value chain and business architecture play a huge role in determining a company's business strategy and how it positions itself in the value chain. Digitization may cause links in a vertically integrated value chain to dissolve. New players may arise who add value in an innovative way to a specific part of the value chain. When platforms are involved, the locus of strength and value capture may shift—from hardware to software or from manufacturing to service provision. When platform dominance is unlikely to be achieved, deconstruction of the value chain may open other opportunities to add and capture value. Perhaps more agility is needed to rapidly innovate new offerings. Plugging into an existing platform may be the best strategy. Many companies have been successful with such a strategy, for example, game-developer Zynga, which created Farmville on Facebook, or Vacasa, which leverages Airbnb to provide vacation rental property management services to homeowners. For established companies, your biggest competitive threat may not be a similarly scaled competitor but dozens of small, specialized competitors capturing specific parts of your value stack.

Consider another example, telecommunications. Telephony was originally offered as a pure fee for service. Dominant providers, such as AT&T, were vertically integrated—providing the wires to your home, operating the switchboard, and even manufacturing the handsets that customers would use as part of their service agreement. Then the rise of the Internet started to deconstruct the value chain. Telephone companies started to face competition from new competitors leveraging voice-over-internet protocols. Cellular networks arose, leveraging the Internet's backbone. Services such as SMS text messaging started proliferating as mobile telephones diffused. In their wake, a series of application-driven start-ups such as Skype and WhatsApp leveraged the new underlying platform architecture to provide services similar to what AT&T used to do in its traditional landline business. New opportunities abounded, from social media applications such as Facebook to value added services such as Apple's Facetime to teleconference applications such as Zoom. AT&T, of course, has survived but only after many iterations, ultimately refining its business to be primarily a cellular service provider.

Such reconfigurations are occurring in other industries. Banking is being disrupted by the rise of the so-called fintech companies. Rather than target the full range of services offered by large commercial banks such as Bank of America or Wells Fargo, the new fintechs typically target a narrow piece of the value chain, leveraging technology to deliver superior value. Whether it is Venmo in peer-to-peer payments or Stripe in payment processing for online businesses or Chime for fee-free online banking, they are eating away at traditional banks and deconstructing the value chain. A similar dynamic may play out in electric utilities. While the backbone infrastructure of wires from large generators to the home will likely remain under the control of utilities, innovations such as smart metering, home monitoring services such as Google's Nest, and the potential for distributed generation such as having solar panels on your home are all capturing new opportunities that will likely reduce demand for traditional electric services and create new market opportunities.

As value chains are deconstructed, modularity as a basis of competition becomes more important. Modularity refers to systems "composed of units that are designed independently but still function as an integrated whole."[4] The key is to define a set of rules about how these independent units work together. Application programming interfaces work this way, defining how a programmer may reference Google Maps in their application or how a newsfeed may include tweets from Twitter. More generally, we are seeing the rise of technology "stacks"—various combinations of technology that form an integrated whole. For digital applications, the technology stack usually includes a set of programming languages, tools, and frameworks that developers use. More broadly, stacks create opportunities for companies to plug into a value chain and offer value in new, compelling ways.

THE NEW FORMS OF VALUE CREATION

So, if digital technologies are helping deconstruct value chains, creating new opportunities to plug into various platforms and deliver value in new ways, what are some of these new ways of delivering value? David Rogers cites five domains—customers, data, value, innovation, and partners—as the critical drivers of value creation in the digital age.[5] Let's start with customers. Traditionally, many businesses' interactions with their customers could be best characterized as episodic, typically when discrete purchases are made. While marketing may target specific types of customers, perhaps those fitting certain demographic profiles, by no means would marketing efforts be able to target an individual customer and all their unique quirks and interests. As a result, product offerings tended to aim to be generally desirable with little individual customization. In such a world, information flows are largely unidirectional, flowing from producer to user as they try to inform and persuade potential customers to buy their product.

Now consider customer engagement in the digital age. Episodic transactions are replaced with a constant stream of network interactions. Customers are engaging on platforms which can track an individual's microbehavior as they consume a product or service, or simply search for a product or service to purchase. Customers are not only consuming information pushed out by companies on their products and services, they are also actively shaping opinions and influencing brands and reputations through social networks. Customers may be creating reciprocal value flows with other customers and with the platform provider. Customers may even see themselves as part of a community of users, assuming ownership of a platform and influencing the direction of innovation and future offerings.

Digitization is allowing for what Nicolaj Siggelkow and Christian Terwiesch call "connected strategies," strategies that turn "occasional, sporadic transactions with customers into long-term continuous relationships."[6] They highlight that connected strategies may simultaneously enhance the user experience while also driving operational efficiencies that may drive costs lower and ultimately lead to lower prices. They share the example of Instacart, an application that allows customers to buy goods online from local retailers. Paid shoppers then go to those stores, purchase the items, and deliver them to the customers at their home in as short as an hour. Like many delivery services, Instacart creates customer value and higher willingness to pay by reducing the time and effort to go and search for goods at a store. However, unlike delivery services focused on one store or customer, Instacart's approach allows for batch processing, in which a paid shopper may combine multiple orders and deliver to multiple homes on the same trip. This, in turn, creates operational efficiencies that lower the fulfillment cost for Instacart with respect to individual store delivery services.

Consider the motion picture industry. Historically, customer engagement was largely unidirectional from movie studio to moviegoer. Studios were able to use box office proceeds, that is, ticket sales, to assess movie success and decide on what new movies to produce. Additional feedback was available through critics' reviews printed in newspapers and the occasional focus group study. The studios had little information on the interests and proclivities of individual consumers. Production decisions were often driven by what seemed to work in the past or what genres seemed to be popular at the moment. Marketing was largely driven by paid advertising in major media outlets such as network television or national magazines.

Enter the digital age. Sentiment aggregators such as Rotten Tomatoes can make or break a film. Movie watchers take to social media outlets such as Facebook and Twitter to share their opinions and reviews of movies, with famous social influencers having an outsized impact on adoption. On occasion, communities of passionate fans may launch campaigns to get new movies made or to alter the trajectory of a beloved franchise. Witness the efforts of fans of DC Comics to get Warner Brothers to release the director's cut of the superhero movie *The Justice League*. For the movie studios, tools like Google Analytics allow them to track search behavior and try to gauge the buzz around a new movie. Using Facebook, they can target hypercustomized advertising to specific individual users. With the rise of movie streaming online, movie producers are now able to collect data on the specific media consumption patterns of individuals.

Such a world—one with continuous, dynamic, two-way communication with your customers—generates a lot of data. Some of this data is structured, data that resides in a fixed field within an individual record that may be stored in a relational database. For example, Netflix collects specific structured data on which movies you start watching, when you watched them, and whether you completed the movie. Some of this data is unstructured, data that does not have a predefined data model or is not organized in a predefined manner. Scraping social media sites for customer buzz on a new movie would be an example of such unstructured data. Several techniques such as natural language processing and image recognition algorithms are making such unstructured data more useful and valuable.

As we discussed in Chapter 1, all this data ultimately provides an opportunity for aggregation—the creation of new sources of value for customers by leveraging the data collected and analyzed. Recall Spotify's use of customer music selections to build prediction machines to make recommendations and curated playlists. Or Amazon's use of previous purchasing history to recommend products to purchase before you search. The data allows for hypercustomization of the user experience. In the predigital era, advertisements in newspapers like the *New York Times* were the same for everyone who purchased a paper (at least within a specific region). In the digital age, an online *New York Times* can customize advertisements to the individual reader on the basis of terabytes of data on that individual. Furthermore, companies can often customize their offerings in real time using continuous flows of new information to update predictions and adjust offerings using this new data.

Siggelkow and Terwiesch propose a four-stage progression of aggregation, from traditional respond-to-desire offerings such as searching for a product on Amazon to curated offerings such as Spotify to what they call "coach behavior," and ultimately automation. "Coach behavior" refers to experiences when predicted algorithms and behavior nudges are used to change customer behavior. Gamification, when customers engage in friendly competition against themselves and peers, is an example. You see this in fitness applications such as Peloton, which prompts users to work out and to encourage other users. Automation refers to automatic execution of transactions with no direct engagement by the customer. Chewy, an online provider of pet products, has a service that estimates when a customer may need more pet food and ships it to their home without any actions on the customer's behalf.

The opportunity to curate, and even automate, offerings in real time on the basis of voluminous user data changes the fundamental nature of the relationship between producer and consumer. In the digital age, your value proposition may be constantly changing to meet ever-changing customer needs. No longer do you have the luxury of simply trying to execute on a stable value proposition. Whereas innovation used to be a discrete activity driven largely by internal intuition and experience, innovation in the digital age is a continuous process of updates and new offerings. The emphasis is on quick and inexpensive learning in which failure is not viewed with derision if you fail fast and pivot.

Platforms often provide an opportunity to run true experiments such as AB testing with users. The hallmark of experiments is the ability to randomly assign a group of participants to a condition. AB testing is a particular type of experiment popular with application developers in which two randomly assigned groups are given different experiences (A and B). Companies such as Facebook, Amazon, Netflix, and Google make great use of AB testing to assess new product features and offerings, running hundreds of experiments a day. They have hired scores of behavioral and data scientists to design and execute these experiments.

Experimentation is not limited to the platform providers. Customers themselves, or users more broadly, may be encouraged to customize offerings and to generate innovations. Lego, the maker of the iconic building block toy, encourages its users to play online with digital Lego bricks, to create and share structures, and to rate other user's creations—the most interesting of which it may manufacture and sell as physical Lego kits. This is an example of what has long been referred to as user-led innovation. Digitization simply makes user-led innovation easier to facilitate and track. In this way, customers become a true partner in the innovation process.

Such partnerships may not be limited to your customers. Other companies in your value chain may become key innovation partners. Amazon encourages those providing goods through Amazon Marketplace to experiment and innovate. Coursera encourages its partner universities to experiment with new offerings and new modes of online delivery. In some cases, the distinction between competitor and partner may blur. While many of the big tech companies are increasingly competing head-to-head with one another, they also continue to rely on each other as partners in value creation. Amazon allows its Alexi voice assistant devices to connect with and control Google Nest products. Apple iPhone users may use Gmail or Google Maps as their primary email provider and mapping application. Together they are engaging with customers and creating customer value.

THE EMERGENCE OF NOVEL BUSINESS MODELS

As digitization creates new ways to create value, it is also driving new ways to *capture* value. We use the term *business model* to refer to the way a business captures value from its offerings to users and customers. Stated another way, the business model tells us how a company generates revenues and from whom. Traditionally, business models were largely bilateral in the sense that a company would sell a good or service to a customer who would pay a predetermined price, such as Apple or Samsung selling you a smartphone, or Microsoft selling you a copy of its Office suite of tools. As the digital age has unfolded, we have seen a proliferation of alternative business models—some old, some new—that are changing the dynamics of competition within industries (see Figure 3.2).

The rise of the Internet coupled with the increase in bandwidth and the buildout of cloud computing has helped proliferate software-as-a service (SaaS) models. Rather than a one-time purchase of a discrete software license—perhaps delivered on a compact disc in the old days or likely downloaded online today, SaaS charges a subscription fee for the use of software over a period of time. For example, you may pay a monthly subscription or perhaps pay as you go. The most obvious impact this has had on software is that rather than having very discrete timed releases of new software updates, now users can be provided updates in real time. The development cycle for software moves from a staged approach to a continuous process of innovation. For businesses, the relationship with the customer changes

| Fee for Product | SAAS | Advertising | Freemium | Partnering |

Figure 3.2. Digital Business Models.

from a constant effort to get them to pay for new upgrades to an effort to maintain a relationship.

The most interesting advance in the digital age has been the explosive growth of advertising-driven models. Rather than charging users a fee, users are given free access to an application or a service with the trade-off of being exposed to advertisements. The service provider ultimately generates revenues from the fees it charges advertisers. Of course, an advertising business model is nothing new. Newspapers, radio, and television had been using advertising since their inception. However, several large tech providers have leveraged an advertising model to become some of the most profitable companies in the world. Over 80 percent of Google's revenues[7] and 98 percent of Facebook's revenues[8] are from advertising. In the case of Google, it gives away a whole suite of software applications from its core search product to Gmail to Google Maps to the Android operating system, in exchange for users having to view advertising.

Central to these business models is data. While users are given free access to applications, in using those applications, they are generating volumes of data that can be used to refine algorithms and target advertising to a specific user. Advertising-driven models in the past, such as television, relied on program characteristics in charging premiums to advertisers looking to target certain demographics. For example, a sitcom such as NBC's *Friends* about young people living in New York City may attract a younger, urban audience that advertisers covet. In the new digital age, advertising has gotten hyperspecific—ads targeted specifically to you on the basis of your behavior across a wide number of digital artifacts. These new ad opportunities are more refined and more valuable, typically leveraging predictive algorithms generated using machine learning approaches. Competitively, this is fueling data-driven network externalities for companies such as Facebook and Google and giving them dominant positions in their marketplaces.

The growth in the advertising business model has given rise to a whole host of related "freemium" business models. Freemium models rely on a two-tiered system in which users can access services for free but pay a premium for additional content or services. For example, the *New York Times* provides much of its daily newspaper online for free. However, users are limited in the number of articles they can read per month and must pay a subscription fee to unlock all the content available. Spotify's music stream service can be used for free, but one is exposed to ads. To remove the ads, one must pay a monthly subscription fee. We are seeing the proliferation of freemium models in the video content space as traditional advertising-driven television increasingly moves behind the paywalls of streaming services such as Disney+ and HBOMax.

Another business model that has grown in popularity in the digital age is what we might refer to as "partnering models." Partnering models are when a facilitator to a transaction, such as a platform, charges a percentage fee for any funds exchanged. For example, Apple charges a 30 percent fee for every app purchased or in-app fee paid for applications available through the Apple App Store. Such fee structures are common for two-sided market makers such as Uber and Airbnb. Technically, customers are paying a fee to the other party of exchange—the driver or house provider, respectively. However, each platform collects a percentage of those fees; around 25 percent of the ride fee in the case of Uber. Airbnb, however, charges both providers and renters, roughly 3 percent and 14 percent, respectively. As we discussed in Chapter 2, some companies choose not to charge such service fees to attract users as they try to establish their platform. However, they will likely eventually do so to monetize their offering once a platform is established.

Sometimes multiple business models may be in play in the same industry. In the battle for mobile operating system dominance, Apple largely uses a traditional fee-for-product model, charging a premium for its iPhones, which come with iOS installed. Google uses an advertising model, giving away its Android operating system to collect user data to refine its targeted advertising. The "best" business model is driven not only by market conditions, but by the specific strategy and capabilities of an organization. Recall the Strategist's Challenge: firms should seek to carve out defensible competitive positions that differentially outcompete rivals at the intersection of their values, capabilities, and market opportunities.

THE EVOLUTION IN COMPETITIVE POSITIONS

In his famous treatise on competition, Harvard Business School professor Michael Porter identified four generic defensible competitive positions available to companies.[9] Modifying slightly, we may refer to these as the cost leader, the quality differentiator, the niche player, and the integrator. The cost leader is typically a company that commands large market share to capture economies of scale to drive down costs and ultimately charge the lowest prices in the market. Think Walmart among US grocery stores. The quality differentiator incurs higher costs to offer differentiated goods and services for which customers are willing to pay a price premium. Think Whole Foods (interestingly, owned by Amazon since 2017). The niche player chooses to compete in a narrower segment of the market, offering specialized goods favored by these target customers. Consider Bueche's Food World, an institution in Flushing, Michigan, that provides a family-friendly community experience. Last, the integrator aims to have the best of all worlds—to offer quality goods and services at a discount to a large market. In grocery, Albertson's arguably is aiming to be an integrator. While such a position is often coveted, it can be incredibly hard to achieve.

These four generic positions remain relevant in the digital age. Look no further than Apple and Samsung. In the smartphone market, Apple is the quality differentiator charging high premiums for its iPhones. Samsung is arguably the cost leader, producing at scale, leveraging the open-sourced Android operating system, and charging prices far below that of Apple's iPhones. Or consider the market for online subscription box clothing services, with which customers receive curated collections of clothes selected according to users' specific preferences (see Figure 3.3).[10] Stitch Fix was a pioneer of the concept and offers a broad selection and competitive pricing. Trunk Club is a competitor that was acquired from Nordstrom and aims to be a quality differentiated player. Dia & Co is a niche player that targets plus-size women. Amazon offers Prime Wardrobe, a "try-before-you-buy" service that arguably aims to be the cost leader.

In the case of online subscription box clothing services, it is interesting

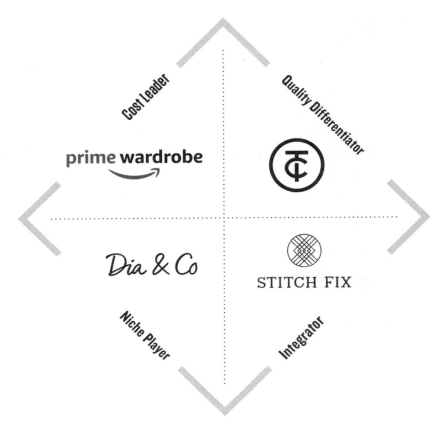

Figure 3.3. Classic Strategy Positions.

to note the role that data plays and how that may have an impact on the viability of different positions. Central to the concept is to leverage data on past behavior to carefully curate a set of recommended clothes. This data includes specific purchases of a buyer as well as the purchases of other buyers who may share similar characteristics or preferences. Using machine learning algorithms, these service providers use this data to develop predictions on what clothes you may like. As we have emphasized throughout this book, such data-driven strategies create increasing returns to scale that can serve as a barrier to competition. Those providers who can acquire larger datasets will have more accurate predictions, attracting more customers and creating a virtuous cycle. All of which leads to the observation that the coveted integrator role may be more likely in a digital world in the presence of network externalities. Stitch Fix, as the early market leader, may be able to leverage its data advantage to achieve an integrator position competing effectively against cost leaders, quality differentiators, and even niche players.

To help us further unpack competitive positioning in the digital age, we can identify some other common positions. Hamilton Helmer in his work "7 Powers" identifies seven classic strategic positions.[11] Those pursuing "Scale Economies" typically aim for broadly desired goods and for early and quick growth to achieve economies of scale and low-cost production. Those pursuing "Cornered Resources" are competing to secure scarce resources, perhaps prime oceanfront property to build a seafood restaurant or a patent on a new drug to cure baldness. Those pursuing a "Branding" position are trying to leverage marketing and a distinct identity to increase consumer willingness to pay. Think Nike in sports apparel. Those pursuing "Processing Power" are looking to leverage learning curves and interdependencies in production to create efficient operations that are difficult for others to copy, such as Toyota's vaunted lean production system.

Each of these four positions remains relevant in the digital age. Apple clearly uses Branding to help build barriers to competition in the luxury smartphone market. Amazon has built an impressive logistics infrastructure that serves as the backbone of its e-commerce effort, leveraging both Scale Economies and Processing Power. Google's patent on rank algorithm search was a Cornered Resource that allowed it to thrive, especially early on in its growth.[12]

Yet it is arguably the last three of Helmer's "powers" that are the most relevant to the digital age: "Switching Costs," "Counter-Positioning," and "Network Economies." A Switching Cost position refers to efforts to lock in customers to a desired technology or platform. Clearly, this is true with operating systems such as Microsoft Windows or Apple iOS. Counter-Positioning refers to efforts to outmaneuver established rivals by offering new products, services, or business models that those rivals are unlikely to match. Perhaps those rivals are worried about cannibalizing existing offerings or diluting their brand, or simply see a counter position as too small or unattractive. As an example, a lot of fintech start-ups are using Counter-Positioning strategies relative to established commercial banks.

Last but not least is the Network Economies position. As we have discussed at length, network externalities are endemic to the digital age. The increasing returns that accrue as more users adopt a particular technology or platform creates a virtuous cycle that makes it hard for competitors to dislodge once established. Most of the big tech companies today have captured some form of network externality. The challenge, of course, is winning the platform battle. As we discussed back in Chapter 2, there are many factors to consider when trying to capture the Network Economies position: whether to be more open or closed, which business model to pursue, the timing of entry. All too often, it is assumed that the only position available is the Network Economies position. This is especially true in the presence of what are feared to be winner-take-all dynamics. The race for platform dominance is assumed to resemble a battle royal with only the last survivor standing in the end.

Fortunately, that is rarely the case. Even in the presence of true winner-take-all dynamics there may be viable positions either in niches or elsewhere in the value chain. While these may not be as lucrative as the Network Economics position, it is arguably far better to strive for achievable positions than a likely doomed attempt at platform dominance. I have come across hundreds of established companies telling themselves that they are going to be the "Google of X," X being whatever market segment they currently play or aspire to play. All too often, Google is more likely going to be the Google of X (or maybe Amazon or Apple). Identifying a valuable competitive position is far different from securing that position. When strategizing, it helps to have a dose of humility and realism.

THE SEARCH FOR VALUABLE POSITIONS

So, what competitive position makes the most sense for *your* organization? To answer this question, let's return to our analysis of capabilities, first mentioned in Chapter 1. What are your current core capabilities as an organization? What are the people, processes, and systems that undergird these capabilities? To what extent are the capabilities you possess well aligned with one another and with your overall value proposition? To what extent could a capability or capabilities provide a true competitive advantage? Would it be hard for others to imitate those capabilities or deliver value in similar ways? How durable is any competitive advantage that you may have?

Consider Apple. When Steve Jobs returned to Apple in 1997, the company was in disarray. After several disastrous moves including experimenting with allowing Mac "clone" computers by other manufacturers and the failed release of the Newton, a digital assistant either well ahead of its time or an example of Apple losing its way, Jobs quickly pivoted Apple back to what he saw as its core capability: innovative designs that delivered an integrated "wow" experience to customers. He reinvigorated his design team, promoting Jony Ive to Senior Vice President of Industrial Design. Their first effort was the new iMac line released in 1998 to rave reviews. However, the company continued to struggle, as it had long lost the Network Economies position to Microsoft and the Windows OS. It simply wasn't profitable to produce a high-end machine when you commanded only 3 percent market share.

So, Jobs focused Apple's design talent in a new direction, towards consumer electronics—eventually releasing the iPod and iPhone. These nascent markets provided Apple a fresh start and a less contested market. It entered early in the Competitive Life Cycle for these products so there was no dominant platform or operating system. After a few early missteps, Apple worked consciously to build an ecosystem of products and services, such as iTunes and eventually the App Store and iCloud, that raised Switching Costs and captured Network Economies. All the time, it stuck with its core capabilities: designing a suite of products that wowed and allowed Apple to position itself as a high-end differentiated player with an unrivaled brand.

There are several lessons to learn from Apple. As highlighted in our Strategist's Challenge, strategy is about understanding where values and capabilities meet ever-evolving market opportunities. Apple was able to capitalize on the emergence of new technologies that basically reset the Competitive Life Cycle, while leveraging its historic strengths in product design, to establish a highly profitable position. Apple through one lens was stable and consistent: it stuck with its core strength in design, pursuing vertically integration marrying hardware and software. Through another lens, Apple was adaptive: recognizing the folly of competing against a well-entrenched platform, it pivoted to more open competitive terrain.

This tension between static and dynamic is central to strategy. Digital transformation is often about how you evolve your set of capabilities. I very carefully use the term *evolve*. Rarely do you see a company successfully reinvent its capabilities wholesale. Blockbuster Video investors saw the folly in transforming a brick-and-mortar chain store into a streaming powerhouse like Netflix. Business professor Dorothy Leonard famously wrote that core capabilities often become core rigidities.[13] What made you successful in the past makes it more difficult to transform your capabilities. We've been successful in the past, why should we change? This is one of the reasons that entrepreneurial ventures are often the winners when new technologies and business opportunities present themselves.

All is not lost, however. Organizations can and do evolve their core capabilities. IBM was able to pivot from being a maker of computer hardware to being a service and software provider emphasizing artificial intelligence. When Blackberry had clearly lost the mobile computing platform battle, it pivoted to software and cybersecurity, becoming an SaaS provider for large companies. For many organizations, digital transformation is exactly about evolving your core capabilities to meet emerging market needs. The critical strategy task is to target a desired and *achievable* competitive position and then evolve your capabilities to meet the needs of that position.

As you think about how to best position yourself in an evolving marketplace, it is helpful to think in dynamic terms. Recall the Competitive Life Cycle. Hamilton Helmer proposes that the stages of the Competitive Life Cycle help dictate which of the seven power positions are available to you. Early in the Competitive Life Cycle, when a technology is nascent and a dominant design has not been established, is when Counter-Positioning to outmaneuver incumbents in the broader industry and Cornered Resources to try to lock in valuable assets before others are most useful. As a new technology or business begins to take off and enters the growth phase, the battle for increasing returns to scale becomes critical—highlighting Scale Economies, Switching Costs, and of course Network Economies. Finally, as the market matures and a shakeout ensues, Branding and Processing Power may become more important as bases of competition.

Another consideration is your organization's maturation. New entrepreneurial ventures have a set of opportunities and challenges before them that may differ significantly from more established, high-growth enterprises. These high-growth enterprises, in turn, face a different set of challenges than those of more mature companies that have achieved scale and have well-built-out capabilities. For example, new ventures tend to be less bureaucratic and may be better able to quickly test new concepts and innovate. Growth enterprises may have abundant access to capital due to favorable market capitalization or accrued cash. Mature companies may be better able to leverage complementary capabilities in production and

logistics and occupy positions requiring process improvement and cost reduction. The key is to identify positions that capitalize at the intersection
of the firm's maturation and the industry's evolution—where the industry
sits in the Competitive Life Cycle.

To help with this critical question, it can be useful to make use of a basic
strategy tool, the strategy map. A strategy map is, very simply, a visual representation of the positioning of various companies in a market. They can
be two dimensional, using the x and y axes to illustrate different attributes
of competition, or three dimensional, by varying the size of the marker.
Comparisons can be made on various attributes of product offerings, critical capabilities, or competitive choices. Often, it is helpful to construct
multiple maps to compare organizations on several different dimensions.

See the strategy map in Figure 3.4 as an example. Firm 1 is positioned
as a niche player offering a high-quality, sporty styling to a small market.
Firm 2 may be positioned as a cost leader, providing products of average
quality and styling, and commanding a large market share. Firms 3 and 4
may be leveraging differentiated positions aiming at specific market segments. In the market for banking services in the United States, we have
large integrated providers such as Bank of America and Wells Fargo and
smaller specialty online producers, such as Chime. Their underlying capabilities, their resulting value proposition, and even the type of customer
they target differ between these two strategic groups.

Strategy maps can begin to provide a window into what strategic positions are achievable for your organization. Remember, just because you
can identify a valuable position does not mean that you are likely to achieve
that position. Who else may be seeking the same position? You may want
to undertake a gap analysis, in which you assess the difference between
your current state and a desired position. Gaps are likely in the face of digital transformation. The challenge is ensuring the gaps are not too large to
be overcome. Rarely do organizations totally reinvent themselves eschew-
ing completely what made them successful in the past. Paramount is the
durability question from our capabilities analysis: Which of your capabilities will continue to be valuable in the future?

To help identify a desired position, it is helpful to articulate the three
W's: we aspire to be Why, for Whom, by doing What. "Why" refers to
your desired value proposition. Why would customers be attracted to your
product or service? What value do you create and for them? "Whom" refers

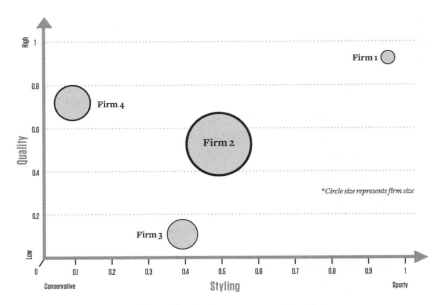

Figure 3.4. Strategy Map. Note: Circle size represents firm size.

to your specific customer base. Whom, specifically, are you targeting with your products and services? What unique attributes and desires do they possess? What channels may you use to reach them? Finally, "What" refers to the specific capabilities you possess to deliver on your value proposition. What key resource, activities, and partners are needed? This may also include your "How," your business model for capturing value—how will you generate revenues and secure a competitive advantage?

The framework at the end of this chapter provides a template for analyzing these questions. The template is well aligned with the Business Model Canvas that is popular among start-ups in Silicon Valley. The Business Model Canvas is a one-page template that includes spaces to articulate your core value proposition (your Why); your targeted customer segments, relationships, and channels (your Whom); your key activities, resources, and partners (your What); and underlying revenue streams and cost structures, in other words, your business model (your How).

FRAMEWORK 3: FORMULATING A DIGITAL POSITION

We begin with our baseline snapshot of the current competitive environment from our first framework and our mapping of the dynamics of competition and the Competitive Life Cycle from our second framework. Where is the organization in its individual life cycle? What capabilities does it possess that can be leveraged in the digital age? Where is the industry in its own evolution? What trends are having an impact on the sector? What is the likely industry archetype to emerge from digital transformation? What about competitors, current and future? What positions may they be seeking in a digital transformed sector?

Use this information to identify a desired future competitive position (see Figure 3.5). Begin with column one: What is your "Why" and for "Whom"? Which of the four generic positions do you envision capturing? Next, consider where in the value chain you will play. For example, do you target an upstream position selling to other businesses (B2B), or perhaps a downstream position selling directly to general customers (B2C). Third, whether B2B or B2C, what are the attributes of the customers you wish to target? Fourth, what is your value proposition to these customers? Your targeted customers and your value proposition should both be well aligned with your generic positioning. Fifth, how are you going to leverage data to deliver on your value proposition? How will this data support the previously identified elements of your competitive position?

Next, turn your attention to your proposed business model (column two). How will you monetize your offerings and generate revenues? Will you choose a traditional fee-for-service model or perhaps leverage a software-as-a-service or advertising business model? Maybe a subscription model or freemium model makes the most sense? Check the box for each business model you plan to use. Use the space remaining in each checked box to provide additional details about your approach. While these business models are not mutually exclusive, recognize that it would be rare for you to choose every one of these. Remember, strategy is as much about articulating what you are *not* going to do as what you plan to do.

Last, in column three, identify how you plan to secure your desired position. Consider the seven powers. Do you try to lever a network externality? Do you try to build out a distinctive brand or raise switching costs? Perhaps you target more traditional cornered resources, like patents, or

Why / Whom	What	How
Generic Strategy What is your generic strategy? Do you aspire to be the cost leader, a quality differentiator, a niche player, or an integrator?	☐ **Fee-for-Product Model** Traditional model in which a fee is paid for a discrete product or service.	☐ **Network Economies** Lever network externalities to offer a unique product or service while building a defensible position.
		☐ **Switching Costs** Attempt to lock in customers to a desired technology or platform by creating switch costs to alternatives.
Value Chain Position Where in the value chain do you play? Upstream B2B or downstream B2C? Both? How vertically integrated are you in the value chain? Who are your partners in the value chain?	☐ **Subscription Model** A model in which a subscription is paid. Software as a service is one example.	
		☐ **Counter-Positioning** Outmaneuver rivals by offering new products, services, or business models that rivals are unlikely to match.
Targeted Customers Who are your target customers? What demographic segments, geographies, etc. are your products and services designed for?	☐ **Advertising Model** A model in which users do not pay a fee but are exposed to advertisements.	☐ **Scale Economies** Aim for early and quick growth to capture economies of scale that discourage competition.
Value Proposition What value do you desire to create for your targeted customers? What unique attributes will attract customers/partners to your products and services?	☐ **Freemium Model** A model in which basic services are free, while premium services require a fee.	☐ **Cornered Resources** Aim to secure scarce resources, such as patents, to establish a valuable defensible position. ☐ **Branding** Leverage marketing and a distinct identity to increase customer willingness to pay for your products and services.
Data Value How can you leverage data to enhance the value proposition? When should offerings be automated to create more data? How can data be used to create long-term continuous relationships?	☐ **Partnering Model** A model in which a partner, such as a platform provider, charges a percentage fee on every transaction.	☐ **Processing Power** Leverage learning-curves and interdependencies to create advantages that are difficult for others to copy.

Figure 3.5. Formulating a Digital Position Framework.

scale economies? Once again, these strategies are not mutually exclusive, though it would be highly unlikely to pursue all at the same time. Your "How" should be consistent with the realities of the market as reflected in the industry archetype from the second worksheet. It should also be achievable in accordance with your desired competitive position and your current set of capabilities. Remember, while new capabilities may need to be developed, ask whether you can realistically do so. This is a theme we will explore in more depth in the next chapter. Our next framework will provide an opportunity for mapping out capability development.

4

Appropriating in the Digital Age

General Electric saw the digital future coming and prepared. The venerable maker of large industrial equipment, GE had been one of the preeminent business leaders of the twentieth century. From locomotives to electric turbines to industrial machinery, General Electric was an innovator and often the market leader. For the oldest company in the Dow Jones stock index, digital transformation at GE meant leveraging the Internet of Things to deliver new sources of value for customers and capturing new revenue streams based on services. In 2012, under Jeff Immelt's leadership, the company announced its Industrial Internet initiative to great fanfare. By embedding sensors into its equipment, it was able to generate massive amounts of data on the performance of its machines. GE was going to create an open, global network that connected complex machines and data to provide predictive analytics for improving operations, identifying efficiencies, and reducing downtimes. In doing so, it was going to transform its business model from selling machines to providing value-added services.

Fast forward to a decade later. The market value of the global "industrial internet of things," a term pioneered by GE, is estimated at over $200 billion and forecasted to grow to over a trillion by 2028.[1] Yet GE's market valuation was only $116 billion despite $75 billion in annual revenues, nearly half of its market value in 2012 when it announced its Industrial Internet initiative. A host of competitors were offering similar services, including a diverse array of new entrants, established IT players such as IBM, Cisco, and Intel, and traditional GE competitor Siemens. Analysts openly questioned whether GE's historic strength in hardware translated into a strength in IoT. The company planned to spin out its digital initiative as a small component of a new energy business. What was clear was that IoT was a huge market opportunity and that GE had not appropriated much of its value to date.

History is littered with organizations that failed to appropriate the gains from innovation. The CT scanner, used in medical imaging, was innovated by EMI in its central research laboratory in 1972, winning its inventor, Godfrey Hounsfield—an electrical engineer at the company—a Noble Prize in Medicine. Yet it was GE that was able to leverage its existing position in X-ray machines to dominate the CT scanner market. Xerox, in its famous Palo Alto Research Center (PARC), had innovated the computer mouse and the graphical user interface back in the 1970s. Yet it was Apple with its Mac (and later Microsoft with Windows) that largely captured the value from these innovations after Steve Jobs famously visited PARC and capitalized on the innovations when Xerox proved unable to see their true value.

Even when companies have the foresight to innovate new impactful digital technologies or envision valuable future competitive positions in the digital age, the question that remains is, How do we capture or appropriate value from our innovations? Arguably, every viable competitive position needs to create value for multiple stakeholders such as customers, employees, and suppliers. Otherwise, why would these stakeholders be partners to exchange? Appropriability refers to the share of value captured by these different parties. This includes competitors who may imitate your efforts or benefit in other ways. The question for most companies is, How do we ensure that we capture a significant amount of the value created by our efforts?

At its heart, the appropriability question is really a question of competitive advantage. Having a bold strategic vision is not enough. Creating value for other stakeholders, such as customers, is absolutely necessary but may not be sufficient. How can you outcompete rivals who may be jockeying for similar competitive positions? What of new entrants or diversifying incumbents who look to copy your efforts? How can you build barriers to competition to protect your position and capture value? What capabilities may prove valuable, rare, and hard to imitate? How do you capture some of the value that you create?

THE FOLLY OF COMPETING ON TECHNOLOGY

One of the common mistakes that companies make in the digital age is assuming that those with the best technology win. In this simple logic, digital transformation is an arms race among organizations to have the best and latest technology. In fact, businesses and organizations spend over $4 trillion a year on technology.[2] Yet it is unclear that all this technology spend is generating results. Nobel Laureate Robert Solow once famously quipped, "You can see the computer age everywhere but in the productivity statistics."[3] He was highlighting the disconnect between IT expenditures and productivity growth—what is now commonly referred to as Solow's Paradox. For example, whereas IT expenditures have increased twentyfold over the past forty years, productivity has only increased threefold.

There are many reasons for this disconnect. For one, the benefits may be delayed. There is increasing evidence that digital technologies *are* driving productivity gains, but in some cases only decades later. The rise of the Internet as a commercial vehicle in the 1990s has led to an increase in labor productivity; however, those results did not substantively materialize until the 2010s.[4] There is limited evidence that the new AI- and data-driven phase of digitization we currently find ourselves in has increased productivity yet. Such a disconnect between future-looking technology optimism and past-looking economic disappointment may very well be a feature of periods of rapid technological change.[5] Thus there is great hope—and quite a bit of worry—that artificial intelligence will reduce the need for labor, increasing productivity while also reducing employment opportunities.

In general, the benefits of technology investment may be diffuse and complex. Diffuse in the sense that most of these benefits may accrue to complementary activities. Within many organizations, this means that the return on IT investments does not materialize in the IT department itself but in other operating units that leverage the technology provided by IT. The path from cause to effect can be quite complex. Processing data in real time and generating fancy data dashboards may seem like a waste of time and money, especially if these tools have no discernable impact on decision making. Yet perhaps the dashboard highlights an opportunity to innovate a new feature or service that ultimately increases sales for your product. Connecting the initial technology investment with this innovative outcome may be difficult.

Last, the benefits of technology investment may be competed away as others make similar investments. While critical, technology spend does not serve as a competitive advantage if others can make similar investments. This is a critical strategy point—competing on valuable resources is incredibly difficult if those resources are widely available, easy to imitate, and/or substitutable in some way. At the heart of this dilemma is what we call the Red Queen Effect in reference to the character of that name in Lewis Carroll's *Through the Looking Glass*, his sequel to *Alice's Adventures in Wonderland*. The Red Queen prophetically laments that she always seems to be running but never going anywhere. She and her minions are literally running in place. The Red Queen Effect highlights the folly of trying to outspend rivals. As you spend on the latest and greatest technology, if competitors do as well, your relative standing does not change. You are, in essence, running in place.

This raises what we refer to as the Fundamental Principal of Business Strategy: in perfectly competitive markets, no firm realizes economic profits, or what are often termed "economic rents." Stated another way: if everyone can do it, it's difficult to create and capture value from it. The definition of an economic profit or rent is returns in excess of what an investor expects to earn from investments of similar risk. In other words, more than the opportunity cost of the capital deployed on the investment or more than the return from any alternative uses you might have for that money. Competing in the digital age is no different. Competitive advantage rarely accrues to those who simply invest in technology. If only it was so simple!

THE CHANGING ROLE OF INTELLECTUAL PROPERTY

If technology spend is not the basis for competitive advantage, what about various forms of intellectual property (IP) protection? Historically, IP in the form of patents and copyrights would provide a legal barrier to imitation and provide a sustainable competitive advantage, at least until patents and copyrights expired. For pharmaceutical companies such as Pfizer to entertainment companies such as Disney, building and securing legally protected IP has been the foundation of their success. IP protections provide a mechanism for shielding innovations from imitation and create incentives to invest heavily in R&D and to advance new technologies and offerings.

Such IP protections can still be the basis for competitive advantage in the digital age. Recall Google's patent on rank algorithm search. Technology companies such as Apple and Intel have huge patent portfolios. In 2020 alone, Apple filed close to two thousand patents.[6] Google/Alphabet was not far behind with seventeen hundred.[7] Facebook and Intel filed eight hundred plus and six hundred plus patents, respectively.[8] Amazon famously patented "one-click" shopping in 1999, using the patent to successfully thwart imitation by rival bookseller Barnes and Noble and extracting licensing fees from Apple when it adopted the technique for its iTunes store.[9] While Amazon's one-click patent expired in September 2017, the advantage it provided in the early days of e-commerce was instrumental to Amazon's success.

Yet there are reasons to be cautious when pursuing an IP strategy in the digital age. While it is certainly possible to receive a patent on software and related business processes, such patents are notoriously hard to defend. Unlike a patent on, say, a drug, when the formula is well specified and it is easy to verify if violated by a competitor, software is much trickier. As every software programmer knows, the specific code could be manipulated in a multitude of ways and still generate the same outcome. As a result, software patents and supposed infringements are often contestable. Enforcing a software patent often requires a bevy of lawyers and a willingness to fight for the long haul. As I often advise my entrepreneurial students, having a patent isn't as important as your ability to defend the patent. For new ventures, even when "in the right" they may be unsuccessful in prohibiting imitation by rivals, especially when those rivals are bigger, are cash rich, and can sustain a legal defense for years. For this reason, economists often refer to software patents as a "weak IP regime."

In the face of a weak IP regime, firms have resorted to several different tactics. Many larger companies engage in a "patent fence" strategy.[10] This refers to building up a large portfolio of patents to serve as a bulwark, or fence-line, against those who may sue for patent infringement on their own patents. In essence, patent fences try to create a mutually assured destruction equilibrium. During the Cold War, the argument for maintaining large nuclear arsenals was to deter rivals from using their nuclear weapons, for if they did, it would result in the mutual annihilation of both parties. Patent fences work in the same way: if you sue us for patent infringement, we will countersue for infringement on our vast portfolios of patents. Such a strategy arguably was behind Google's 2012 purchase of Motorola. The patents that Google acquired from Motorola gave them a defensive mechanism to fight Apple and others who were trying to sue them for patent infringement. Not that such mutually assured destruction always works. Witness Apple's and Samsung's epic legal battle in smartphones. After seven years, they finally settled out of court for an undisclosed amount.[11]

Of course, legal protections like patents and copyrights are not the only way to defend intellectual property. Firms may be able to leverage other forms of first-mover advantages to capitalize on their innovations. First-mover advantages are simply advantages that accrue to being an early entrant into a nascent market or an early provider of a new technology. See Figure 4.1 for some examples. Recall the "7 Powers" referenced in the previous chapter. Patents and copyrights are examples of a Cornered Resource strategy. If you can scale quickly, you may be able to capture Scale Economies before rivals, achieving lower costs and capturing further market share. If Branding is important, early entry may help build higher willingness to pay among customers attracted to a definitive brand. If learning curves are important, a firm may be able to leverage Processing Power to stay ahead of rivals and continue to dominate the market.

Of course, each of these strategies has their limits in the digital age. For software, the ease of scaling often attenuates Scale Economies. MySpace's early lead in social media was quickly overcome by Facebook, which required little capital investment to achieve similar scale. As for Branding, history is filled with iconic brands left in the wake of upstart rivals. Today, I may make a "Xerox" on my Canon copier. Before Google, there was Yahoo, Lycos, and AltaVista. Even Processing Power attributed to learning curves may be short-lived in a hypercompetitive environment marked by the rapid evolution of new disruptive technologies.

As discussed in Chapter 3, in the digital age, first-mover advantages may most likely be attributed to Switching Costs and Network Economies. In the former, early adoption may lock in customers if they face significant costs, in either effort or money, to switch to an alternative product

Examples	Advantage Type
Secure patents or copyrights	Cornered resource
Build scale quickly	Scale economies
Establish a pioneering brand	Branding
Move down the learning curve	Processing power
Lock in early adopters	Switching costs
Leverage data to enhance offerings	Network economies

Figure 4.1. Types of First-Mover Advantages.

or platform. Consider switching from the Apple iOS to Google's Android or vice versa. In the latter, early adoption may increase the value proposition of the product or platform, leading to a virtuous cycle attracting more customers and further enhancing the offering's lead over rivals. As we have highlighted several times, data may provide a positive network externality that helps defend a position from imitating rivals.

THE IMPORTANCE OF COMPLEMENTARY ASSETS

In the digital age, traditional forms of intellectual property protection, such as patents and copyrights, may not be the strongest basis for competitive advantage. Alternative forms of first-mover advantages such as creating Switching Costs and capturing Network Economies may be more fruitful. Yet this still raises the question of how to beat out rivals jockeying for similar positions at roughly the same time hoping to capture similar advantages. Once again, simply being able to envision a valuable competitive position, even one consistent with current or developing capabilities, may not be sufficient. What complementary capabilities or resources may be necessary to outcompete rivals and achieve your desired position?

Complementary assets have long been viewed as key to securing valuable contested positions in markets. Complementary assets, or simply complementarities, refers to the broader set of resources or capabilities needed to successfully commercialize an innovation, beyond the core innovation itself.[12] In traditional product markets, this might include manufacturing capabilities or a dedicated sales force. Consider a biotech start-up. While a patent on a new drug may provide a high degree of IP protection and thwart imitators, the biotech company will not be able to capture value from the innovation unless it finds a way to manufacture and sell the drug. Does the company develop these capabilities itself or does it rely on partners to provide the necessary complementarities, perhaps an established pharmaceutical company? If the latter, how much value from the innovation does the partner ultimately capture relative to the innovator?

At the heart of complementary assets is the concept of interdependency—the dependence of two or more activities on each other.[13] Stated another way, the value of one activity or asset is dependent on another. We say that two or more interdependent activities are complementary when their joint adoption leads to an increase in value or, alternatively, when the marginal value of engaging in each one is increased by engaging in the other. Complementary interdependency is key to such things as lean production, when only through the joint adoption of a whole suite of activities—just-in-time manufacturing, quality work circles, continuous improvement—can you create maximum value. As a result, the ability to capture value from an innovation relies not only on the strength of IP protection such as patents but on who controls scarce complementary assets necessary to fully commercialize the innovation.

Consider our discussion in Chapter 2 about establishing a dominant platform and "the chicken and the egg" problem. Stakeholders are attracted to platforms that have many participants. To attract participants in the first place then, platform providers often must share all the value with users and partners—in other words, offer their service for free. Whether they can capture value later depends on who holds critical, scarce complementary assets. Recall IBM in essence "sharing" the value of the Wintel standard with clone computer manufacturers such as HP and Compaq. While their open approach allowed their desired platform to dominate, value was largely captured by two complementary asset providers, Microsoft and Intel, who provided the critical and scarce operating system and microprocessor.

In the digital age, potential complementary assets come in many forms and sizes. Brynjolfsson, Jin, and McElheran find in a large sample of US manufacturing establishments that nontrivial returns to the adoption of predictive analytics depend "almost entirely on the presence of other tangible and intangible firm investments."[14] Tangible investments include underlying digital infrastructure—sensors, computing power, bandwidth, cloud storage capacity. Though as I pointed out earlier, such technology spend may be a necessary but not sufficient condition for competitive advantage, especially as much of the needed digital infrastructure these days is easily purchased through cloud providers such as Amazon AWS and Microsoft Azure. That turns our attention to intangibles: things such as human capital, organizational design, and accumulated learning on how to leverage predictive analytics in organizational decision making and judgment.

Thus the presence of complementary *intangible* assets may very well be the key to competitive success. For Amazon, its ability to create an organizational culture that is data-driven and relentlessly focused on innovation and improvement—some may say to a fault—is arguably the secret to its success. For Apple, its expertise in design seemingly undergirds its entire strategy. Apple's investment in human capital, such as hiring leading product designers like Jony Ive, and in organizational culture, creating a dedicated focus on innovating products that "wow" customers, allowed it to forge a position as the quality leader that commands a high price premium on its products.

Brynjolfsson, Jin, and McElheran point out that predictive analytics are dependent on historic and current data to forecast future outcomes.[15] The creation of unique and robust datasets to feed predictive algorithms may very well be the defining competitive advantage in the digital age. The creation of these data reserves is often dependent on prior decisions made about how to organize production within the firm. Ironically, more stable processes better lend themselves to predictive analytics, since the past can be a good predictor of the future. For example, continuous-flow work processes—such as chemical manufacturing or loan processing—tend to be more stable, facilitating the automation of data collection and creating opportunities for continuous improvement.

This raises a classic tension between what is referred to as *exploitation* versus *exploration*. Exploitation refers to optimizing the creation of existing products and technologies by focusing on process improvements and minor changes in offerings. Exploration refers to larger-scale innovations that may disrupt existing offerings, delivering value in new and impactful ways. Most companies intuitively know that they must balance exploitation and exploration. Too much exploitation and you risk being left behind as competitors move the market forward with new innovative offerings. Too much exploration and you risk spinning your wheels, always looking for the next opportunity without capturing sufficient value here and now to have the resources to continue to invest in the future.

Remember our classic technology S-curve. Early in its evolution, R&D and investment in innovating a new technology is often inefficient—lots of time, effort, and money chasing various paths that may never bear fruit. This is exploration. Later, as the technology matures, innovations are harder to come by. Time and money are spent eking out modest improvements. This is exploitation. While it is tempting to simply focus on the steep part of the S-curve, these moments are transitory, and one needs to have a long-term, dynamic perspective. Balancing a portfolio of exploration and exploitation helps you optimally time technology evolution and industry dynamics.

Consider predictive analytics once again. While predictive analytics may align better with stable work processes, that may also be less likely to serve as a long-term competitive advantage in such environments. The cumulative value of data is likely to display decreasing returns, providing an opportunity for competitors to in essence catch up as they copy leaders. Stated another way, the technology S-curve may be rather compressed, quickly approaching the tail end of the curve. On the other hand, less structured and dynamic processes likely create more opportunities for data to help reveal new innovations; in essence, to explore new S-curves. In the extreme, data may create increasing returns to scale as it sends companies down a path of constant innovation that is hard for others to catch up.

Consider the tech giants such as Amazon, Apple, Google/Alphabet, and Facebook/Meta. Each of them uses predictive analytics to continuously innovate their offerings, and not just in incremental ways. They are using analytics to explore more radical innovations and new opportunities, especially in adjacent markets. Witness the entry of each into digital payments. Or their entry into broad sectors such as health care and education. This leveraging of data to innovate is key to the dominance of big tech companies and has, in part, given rise to fears of a winner-take-all economy in which these companies do not just dominate specific sectors, but dominate a wide swath of industries, constantly innovating, keeping would-be competitors at bay.

This fundamental tension created by the innovative potential of data has prompted some strategists to suggest that pursuing sustainable competitive advantage is folly in the digital age. All advantages are temporary as markets are constantly upheaved by new disruptive innovations. Better to think in terms of dynamics and not worry about barriers to competition, this line of thinking goes. I, for one, believe this is more of a semantic issue. There is a long stream of work in strategy that highlights the importance of dynamic capabilities—"the ability to integrate, build, and reconfigure internal competences to address, or in some cases to bring about, changes in the business environment."[16] In essence, the key sustainable competitive advantage may very well be the ability to innovate.

THE INNOVATION IMPERATIVE

The digital age has spawned a dynamic, hypercompetitive environment that places a primacy on innovation. Of course, the quest for innovation is nothing new. Austrian economist Joseph Schumpeter was talking about the "gales of creative destruction" nearly a century ago. To Schumpeter, the ability of firms to invent anew was a hallmark of market-based economies and explained their dynamism and success in improving human welfare. Since the dawn of civilization and the rise of markets, there have been efforts to do it better, cheaper, faster.

In the mid-twentieth century, innovation among big corporate players largely focused on building centralized research and development facilities staffed with the best and the brightest scientists. In the United States, ATT's Bell Labs, Xerox PARC, and IBM's Watson Research Center became synonymous with innovation and symbols of America's technological prowess. They spawned many of the foundational technologies of the digital age, such as the transistor (Bell Labs, 1947), dynamic random-access memory (IBM, 1966), and the ethernet (PARC, 1973). They invested in both basic and applied research and became fountainheads of new patents and even Nobel Prizes (Bell Labs, over twenty-six thousand patents and eleven Nobels,[17] IBM over thirty-eight thousand patents and six Nobels[18]). In general, these research centers were given great latitude to explore new technologies. The strategy, as I like to say, was to hire some really smart people, place them on a hill, and leave them alone and let them do their thing. Commercialization and monetization were secondary concerns to be figured out later and largely centered around securing patents and vertically integrating into production at scale post innovation.

As we approached the twenty-first century, the nature of corporate R&D began to evolve. Market pressures caused many firms to rethink their commitment to basic research. Pressure mounted to demonstrate potential commercial viability and financial returns early in the development process. Meanwhile, the fundamental technological challenges faced in digital technologies were becoming more complex. It was becoming harder for a given lab to have all the requisite expertise necessary under one roof. This was especially true in software, for which innovations were often cumulative, building off what came immediately before in the broader market space. The closed-door, secretive approach to innovation was proving limiting.

These challenges led to the evolution of corporate R&D to what has been referred to as the "open innovation" model. Open innovation is defined by Hank Chesbrough as "the use of purposive inflows and outflows of knowledge to accelerate internal innovation, and expand the markets for external use of innovation, respectively."[19] In this new model, firms leverage an interconnected ecosystem of innovators both within and beyond the firm. Cisco Systems pioneered an "A&D" approach in which they used acquisitions of upstart technology companies to build out their knowledge base and innovate their product offerings, targeting ventures with patents and human capital rather than market share and profits. Intel is a leader in corporate venture capital, or CVC, when companies act like venture capitalists taking equity stakes in privately head ventures. Unlike traditional VCs, however, they are investing for a window on novel technology rather than purely a narrow financial return on investment.[20]

The shift to an open innovation model fundamentally changed the innovation decision problem faced by managers. In the traditional model, innovation was largely a question of how much to spend and on what projects? In the open model, companies had to decide among a portfolio of mechanisms to advance their innovative efforts: internal R&D, alliances, CVC, licensing, university partnerships, acquisitions. Consider GM's pursuit of autonomous vehicles. Historically, this would have entailed funding its central R&D center in Warren, Michigan. GM recognized that it did not have the in-house expertise needed and acquiring that talent could be time-consuming and expensive. So instead it went out and bought an AV start-up, Cruise Automation, in 2016. GM continued to seek outside expertise, acquiring Voyage in 2021.[21] In one of my favorite examples, Uber went out and simply hired away several of the leading scientists from Carnegie Mellon University's National Robotics Engineering Center. I don't know for certain, but I suspect Uber was able to pay substantially more for talent than CMU.

Of course, external acquisition of talent and expertise is not a panacea for innovation. The new open model has fostered markets for innovation—various exchanges for buying and licensing technology such as online platforms Flintbox and Tynex. These markets foster competition, raising the prospect that the cost of acquiring that knowledge may exceed the value that it brings to the company. Critical to success is investing in what my former colleagues Wes Cohen and Dan Levinthal have called "absorptive capacity." Absorptive capacity is "a firm's ability to recognize the value of new information, assimilate it, and apply it to commercial ends."[22] In the pursuit of an open innovation approach, it is critical that a firm maintain a certain level of in-house expertise, if for no other reason than to be able to be aware of and value potential external knowledge transfer opportunities. Having some of those smart people on the hill can be critical even in an open innovation world.

This may especially be true in the digital age. Further complicating the innovation imperative is the potential for AI-related approaches to automate the innovation process. In the life sciences, AlphaFold from Google's DeepMind AI initiative generates predictions on the three-dimensional shape of proteins from its amino-acid sequence, allowing for substantially quicker drug discovery. Researchers from MIT and Tsinghua University used AlphaFold to help develop antibodies to resist COVID-19.[23] The future may be one in which humans and machines work together to advance science and innovations. Garry Kasparov, the chess grandmaster, comments, "We will increasingly become managers of algorithms and use them to boost our creative output—our adventuresome souls."[24]

THE RETURN OF THE STRATEGIST'S CHALLENGE

Innovation may be imperative, but one's innovation strategy is contingent on one's situation. To be clear, one size does not fit all. Recall our Strategist's Challenge from the first chapter. The key to strategy, even in the digital age, is to identify and secure valuable competitive positions at the intersection of your values, the opportunities provided by the market, and your organization's unique and hard-to-imitate capabilities. The digital age highlights that this is not a static proposition, markets are constantly evolving, and, consequently, you need to be constantly innovating, adjusting your capabilities and, ultimately, your product and service offerings to meet these changing market needs.

Our discussion on appropriability further emphasizes that innovation does not occur in a vacuum. Innovation emerges out of a broader ecosystem of stakeholders including customers, suppliers, partners, and even competitors. The key is to position yourself to leverage your unique complementary capabilities and resources to ensure that you capture some of the innovation's value. However, narrowly focusing on capturing as much of the gains from innovation for yourself as possible can be self-defeating, as it may discourage partners and stifle the very innovation necessary to continue to flourish. Often it ultimately pays to expand the pie. Balance is key. Value creation and capture are both critical and largely complementary.

In the digital age, successful organizations nimbly adjust to an ever-changing competitive environment. They focus on building a portfolio of innovation projects, balancing short-term and long-term, top-line and bottom-line growth. Some innovation projects are about improving existing products and services. Some innovation projects are about scaling your current offerings within the market. Many innovation projects are about diversifying your offerings either by introducing new technologies or pioneering adjacent markets. Some may very well be to disrupt the status quo and foster a new competitive ordering, replacing existing technologies and upending the market.

Rare is the firm that is good at all these types of innovation. What types of innovations flow from your specific capabilities? If you are particularly strong in the fundamental science or technology and are more likely to come up with novel disruptive innovations, then a licensing or partnering strategy may make the most sense, especially if you lack other critical complementary capabilities. Perhaps your organization has a unique capability in connecting with the customer. You may be able to leverage that to build a broader set of innovation partners, adding value by facilitating the bilateral flow of data between your customers and partners. You may not technically innovate yourself, but you are creating a platform on which innovations may emerge, perhaps literally a digital platform. Proctor & Gamble has done that in the consumer product space, creating online mechanisms for user-driven innovation in new products by partners, an extension of what it refers to as its connect and develop (C&D) approach.

To help simplify the innovation decision problem, let's highlight three broad approaches: build, buy, and buddy. *Build* includes the classic approach of funding internal R&D as well as broader efforts to build an innovative organization (something we will tackle in depth in Chapter 5). *Buy* refers to acquisitions of other companies as well as licensing and purchasing technology outright. *Buddy* refers to alliances and partnerships of various types including formal joint ventures and corporate venture capital programs. In all three cases, it is important to recognize that innovation is not costless; it is an investment like any other capital expenditure. It is also important to recognize that innovation is inherently a probabilistic endeavor. There are no guarantees that a given investment will bear fruit. Thus there is value in having multiple shots on goal, to use a sports metaphor.

The primary question is, What mix of build, buy, and buddy investments increases the likelihood of valuable new technologies, products, and services emerging from the broader innovation ecosystem? You will note that I do not specify who brings the innovation to market. It may be your organization. Or it may be a partner in your innovation ecosystem. You can still appropriate some of the innovation's value even if you are not commercializing the innovation directly. In fact, some companies such as Qualcomm have positioned themselves as IP providers, innovating new technology and licensing it to others. This relates back to our discussion of complementary assets. What complementary resources or capabilities do you possess? What do you uniquely contribute to the innovative ecosystem? One need not "invent" to capture the gains from innovation. Appropriability is all about fostering a broader ecosystem of innovation while simultaneously being smart about how you may be able to capture some of the value created.

In the digital age, we are awash in platforms, network externalities, and increasing returns to data. These collectively create both opportunities and challenges to devising and executing robust strategies. What is clear is that you must simultaneously think about your plan to create and capture value, recognizing that such plans are not set in stone. World War II US General, and eventual President, Dwight Eisenhower once famously said "strategy is useless, but strategic planning is invaluable."[25] On the battlefield, the best-laid plans are often rendered meaningless as conditions rapidly change in the fog of war. However, having engaged in extensive planning improves rapid assessment of changing dynamics and facilitates pivoting to more fruitful courses of action. Business strategy also needs to be dynamic, adjusting to the ever-changing realities of the marketplace.

My colleague Scott Snell has a useful metaphor about strategy.[26] He likens it to mogul skiing. I am not much of a skier, but apparently to successfully ski moguls, you need to pick a line—a destination you are aiming for. However, as you descend the slope, you are constantly adjusting and changing your approach in response to the little hills and bumps that define the mogul terrain. If you lower your head, you are likely to fall. While your line may change, you need to keep your head up. Having a line is critical to keep you focused and moving forward. Similarly, one needs a strategy, and the strategy needs to be dynamic. The direction may change but having a direction at any moment is critical.

FRAMEWORK 4: SPECIFYING HOW TO CAPTURE VALUE

In our previous framework from Chapter 3, we articulated an aspired competitive position by identifying our Why, for Whom, by doing What and How. This framework builds on the previous one, providing a template to identify your specific approach to developing your offerings (see Figure 4.2). Begin with your core offerings: those products, services, and innovations that you plan to deliver to establish your competitive position. Describe those offerings in the first-row boxes. How will you develop those offerings? Will you develop them in-house, leveraging internal R&D and product development teams? Perhaps you will buy licenses on the underlying technology or outright acquire another player in this space. Or perhaps you will partner with another company sharing development costs.

Value Chain Components	Build	Buy	Buddy	Appropriation Strategy
Core Offerings Start with the core innovation, product, or service that you plan to deliver. How will you create this offering?	☐ Leveraging R&D and organization innovation and product development efforts to internally create the offering.	☐ Purchasing all or part of the offering from another player. May include licensing technology, corporate venture capital, or acquiring a company.	☐ Partnering with another organization such as forming an R&D alliance or a joint venture.	How do you plan to capture value from your offering? Leverage IP such as patents or first mover strategies such as learning curves or network externalities?
Upstream Needs Consider upstream needs, such as supply and components. How will you secure these?	☐ See above.	☐ See above.	☐ See above.	How do these play in your appropriation strategy? Are they a potential source of competitive advantage? Do they put you at risk of misappropriation?
Production Needs Consider production needs, such as manufacturing or complements like software. How will you facilitate these?	☐ See above.	☐ See above.	☐ See above.	☐ See above.
Downstream Needs Consider downstream needs, such as distribution, sales channels, and service. How will you deliver on these?	☐ See above.	☐ See above.	☐ See above.	☐ See above.

Figure 4.2. Specifying How to Capture Value Framework.

Note that these three options are not mutually exclusive. You may combine these options for a given need. In each box, in columns two through four, fill out your specific approach. Next, fill out column five: How will your organization appropriate value, or at least avoid misappropriation by others, for each offering? Perhaps you will leverage intellectual property such as patents to deter imitation. Maybe you will look to scale quickly to capture scale economies that allow you to maintain a competitive advantage. Your individual appropriation strategies should be consistent with your previously articulated "How" from the previous worksheet.

Next, consider the rest of the value chain. What complementary assets and capabilities are required to deliver your offering? For each one, do you plan to build, buy, or buddy? How do they fit into your appropriation strategy? Rows are provided for upstream, production, and downstream needs. For example, what components or software are required? Do you need to access manufacturing capabilities? How will you distribute your product or service? Pay particular attention to the role of data. What data is needed to deliver on your value proposition (see Framework 3)? Who possesses this data and how will you be able to access and leverage it yourself?

Consider, as an example, a strategy to provide a luxury limousine service to high-net-worth individuals using autonomous vehicles. You plan to monetize your offering using a subscription model and hope to leverage your brand and switching costs to establish a defensible competitive position. How do you develop your core offering, the autonomous limo? Do you innovate in-house, building out your autonomous capabilities, like Google with Waymo? Do you purchase an entrepreneurial start-up to jumpstart your efforts, like GM with Cruise? Or do you partner with another company to jointly develop it, like Audi and Aurora? Regardless of which approach you choose, how do you ensure that you can defend your offering from competitors? In other words, how do you plan to appropriate value? Given your hope to leverage your brand to establish and protect your competitive position, perhaps you worry less about others offering similar vehicles and instead focus on protecting complementary capabilities such as superior service to outcompete rivals. Of course, how do you plan to develop and defend a capability for superior service?

PART II

HOW DO I DIGITALLY TRANSFORM MY ORGANIZATION TO FLOURISH IN THE DIGITAL AGE?

5

Leading in the Digital Age

In 2016, I was asked by my dean to serve as the Senior Associate Dean and Chief Strategy Officer for the Darden School of Business. One of my primary tasks was to lead Darden's very own digital transformation. The need was obvious. Higher education was facing its own digital disruption headwinds. The proliferation of massively open online courses was raising alarms. Traditional, residential-based colleges were under attack for being too expensive and too out of touch with the needs of learners. Even among traditional institutions, the battle for top talent was increasingly migrating to a digital battleground in which data and analytics were being used to identify, recruit, and enroll desired students. More generally, digital transformation was being touted as a way to improve operational efficiency, lower costs, and deliver value-added services to students, faculty, and staff.

Here I was—an academic, a professor—being asked to make real what I had been teaching and writing. While I had a background in consulting and had worked with dozens of companies on their digital transformations, this was new territory for me. I came in with big ambitions. We were going to set up a virtual strategy "war room" where we could monitor in real time a whole host of data on our operations and outcomes and compare them with our peers. We were prepared to adopt the leading approaches in machine learning and artificial intelligence to analyze the vast datasets available to us. I envisioned the analyses we could run, forecasting which types of prospective students would thrive at Darden and lead careers of impact and purpose. I was particularly excited about our ability to discover "diamonds in the rough," underplaced talent who given the right opportunity and environment could flourish.

And then reality set in. Even the most basic tasks proved challenging. Our data resided in numerous pockets, often in spreadsheets on the personal computers of staff leaders. Data was often a source of power in the organization, and those who possessed it were reluctant to share it with others. Where formal database systems existed, they were often islands, unable to speak to and connect to other relevant datasets. Tracking an individual from prospective to admitted applicant to enrolled student to job seeker and eventual alumnus was a herculean effort requiring linking a half-dozen distinct systems, each with unique identifiers. Even when successful in creating valuable digital tools, everything from simple dashboards to sophisticated prediction algorithms, we faced the challenge of getting decision makers to use the digital tools we created.

Meanwhile, digital disruption remained a seemingly distant and remote threat to many faculty. Clearly, online education was inferior. A passing fad perhaps. No need for us to fundamentally rethink about how digitization may have an impact on how we create and deliver value. Then the COVID crisis hit. Within a matter of weeks, we had to pivot to 100 percent online delivery. We were not alone. Suddenly, this old bureaucratic industry was being forced to radically accelerate digital transformation. Literally overnight, faculty who promised never to teach online were scrambling to master the dynamics of synchronous and asynchronous online delivery. We learned quickly that the digital world created both challenges and opportunities. We could deliver value in new ways, even connecting with a new set of students. There could even be new business models to help finance and support higher education.

We have by no means completed our digital transformation at my school. We have made great progress, but much remains to be done. Digital transformation is less a destination than a journey. I take great solace in the fact that my experiences have not been unique—not in higher education nor in the broader world of organizations. Almost every company and manager that I have consulted with has shared similar stories of struggle with the most mundane of digital transformation tasks. Even simple data wrangling and sharing can require a heroic effort.

In this chapter, we will focus on leading in the digital age; how to move from strategy formulation to strategy execution, if you will. Central to our framing of digital leadership and management is what I like to refer to as the Digital Transformation Stack—the set of tools and activities needed to manage and lead a digital transformation (see Figure 5.1). At the base of the stack, we have digital infrastructure—the various technologies and applications that allow you to control and transform your data. Next, we have data analytics that allow you to analyze your data and create value-adding predictions. On top of this, we have digital applications—function-based uses of data and digitization such as for marketing or human resources. Last, we have digital strategies that overlay and integrate the lower levels of the Digital Transformation Stack.

THE BUILDING OF DIGITAL INFRASTRUCTURE

The base of our Digital Transformation Stack is your digital infrastructure. As we discussed in Chapter 4, digital infrastructure is rarely a source of competitive advantage. Yet it is critical none-the-less—a necessary, but not sufficient, driver of competitive success. It is the foundation on which your digital capabilities are built and can facilitate higher-order integrative capabilities that may provide competitive advantage. At the heart of your digital infrastructure is the capacity to collect and process data. This includes inspecting, cleaning, and transforming data into useful assets that can be analyzed to aid decision making and to make predictions.

A common practice in many organizations is to begin with the creation of a "data lake," a central hub for storing and retrieving existing data. The data may be raw and unstructured, with purpose not yet well defined. The data may be very structured, collected with a particular aim in mind, and well specified to create a database with clear fields and identifiers. A data warehouse is an older but similar concept to a data lake and is generally limited to more structured datasets. Regardless of terminology, the important task is to pool your data resources in a central location for retrieval and analysis.

This is not a trivial task. Many organizations stumble during this first step of digital transformation. Data often exists in multiple systems spread throughout the organization. In companies that have gone through several mergers and acquisitions, there may be any number of legacy systems that do not easily connect with other sources of data. Data can become

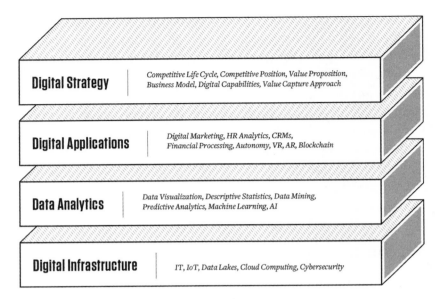

Figure 5.1. The Digital Transformation Stack.

a source of power for those who possess it, and they may be reluctant to share that power with a centralized hub. Leadership is critical to motivate and inspire active participation in the creation of a data lake. Individuals need to understand how they may benefit from such centralization and be incentivized to contribute.

Equally challenging can be integrating data once pulled into the lake. Different identifiers, or no identifiers at all, may make it difficult to connect valuable data from multiple sources. For example, maybe you have robust data on customers from both your sales team and your service team, but the two teams use different ways to identify the same customer. Often the creation of a data lake jump-starts an iterative process for improving and connecting data. You may begin to identify new data items to collect or new ways to record data. This creates a virtuous cycle of data collection, data processing, and data improvement.

For many organizations, this virtuous cycle is leading to identification of new sources of data and new ways of collecting that data. The use of sensors and the broader Internet of Things is creating opportunities to collect data from physical systems. For manufacturers, this may include placing sensors in production equipment or using RFID technology to track supply chains. For retailers, this may include using cameras and sensors to track customer traffic flow in stores and to observe how individuals engage with specific goods for sale. For online and mobile application providers, this might entail collecting a vast array of behavioral data as users engage with the application.

What is clear is that wrangling all this data is critical to your digital transformation efforts. To help facilitate data integration, many organizations are relying on cloud service solutions such as Amazon's AWS and Microsoft's Azure. Cloud computing gives users a relatively efficient and scalable solution for storing and accessing their data and running applications anywhere in the world. Yet the decision to leverage the cloud is not a foregone conclusion for every company. The cloud can present trade-offs worth considering. Connectivity challenges could cause latency in the cloud. Internal legacy systems may be more than capable to handle your data needs and save you money versus subscribing to a cloud service. The cloud may not offer the best level of security for your data and proprietary systems.

Overall, how best to protect the privacy and integrity of your data is one of the critical questions to answer when building out your digital infrastructure. Cybersecurity was close to a $180 billion business globally in 2021.[1] Cybersecurity is the practice of protecting systems, networks, and programs from digital attacks such as hacking to access valuable private data or malware that seeks to destroy sensitive data and interrupt business processes.[2] In 2020, digital extortion through cyberattacks, so-called ransomware, was estimated to cost organizations over $300,000 per attack on average.[3]

All of this highlights the importance of investing in and supporting your information technology (IT) organization. The IT organization is typically on the front lines of building out your digital infrastructure. They are making critical decisions on the underlying technology to store and process your data and how to best protect and preserve that data. These decisions are strategic in nature and require the attention of senior leaders. Long gone are the days when IT spend was merely viewed as a simple capital expenditure if it ever was. The old Chief Technology Officer role has evolved into the Chief Digital Officer in many organizations to reflect the rise in importance of data and digitization. Critical, however, is the recognition that digital transformation is not limited to your IT organization.

They are certainly a catalyst, but digital transformation requires a much broader array of digital champions.

THE RECRUITING OF DIGITAL CHAMPIONS

The second layer of our Digital Transformation Stack is what we refer to as data analytics—the tools and technologies that allow you to analyze data and create value-adding predictions. This includes tools that facilitate the creation of fancy data visualizations—dynamic tables and figures that quickly summarize and communicate data to decision makers. This also includes traditional analytics techniques for summarizing and describing data, including basic statistics and various modeling approaches such as regressions analysis and simulations. Finally, this includes higher-order analytical techniques to facilitate predictions in the presence of large data-sets such as data mining, machine learning algorithms, and random decision forests.

To help organize the vast set of data analytic approaches available, it is useful to consider what each tool or approach helps provide. Descriptive analytics is a broad class of techniques for summarizing and describing what has happened in the past. Simple descriptive statistics, such as the mean and variance, fall into this category. Diagnostic analytics are useful to uncover why something happened, for example, why a machine failed or a package was delayed. Predictive analytics move from describing the past to predicting the future and include everything from regression analysis to basic machine learning. Prescriptive analytics go one step further and suggest a course of action. Autonomy and artificial intelligence fall into this category.[4]

While it is beyond the scope of this book to provide a primer on data science, there are fortunately numerous resources available for learning about and keeping up with the latest advances in data analytics. Software tools such as Tableau, Qlik, and Microsoft's Business Intelligence provide general purpose solutions for processing data and creating compelling data dashboards and other visualizations. Statistical programming languages and applications such as R and Stata were created with the data scientist in mind. Cloud service providers are incorporating advance machine learning algorithms into their suite of offerings, allowing any organization or business access to state-of-the-art analytic approaches.

Of course, the ubiquity of data analytic services also means that core analytics technology is unlikely to be a source of differentiable competitive advantage. This is leading many organizations to place their bets on talent, to seek skilled data analysts to drive their digital transformation. Universities and online educators have responded with a massive increase in data science offerings, with dozens of schools per year offering new master's degrees in data science programs and, in some cases, creating new schools of data science (such as my home institution, the University of Virginia). "Data Scientist" is one of the fastest-growing job opportunities and has been ranked as the top job opportunity by Glassdoor.[5] Data scientists routinely command six-figure salaries and report some of the highest job satisfaction scores.[6]

Every business knows that talent is destiny, but the battle for data scientists can be intense. In a perfectly competitive labor market, talent is unlikely a source of competitive advantage, as wages are competed up to the value the employee produces. Of course, labor markets are rarely perfectly competitive. Smart companies know they must invest in identifying and attracting top data science and tech talent. Established players such as GE and GM emphasize that they're as much software companies as hardware. Even established tech companies such as Google fear looking old and stodgy, worried that new tech talent will see them as an undesirable destination. As discussed in Chapter 4, complementary assets—such as corporate culture—may be critical to recruit the digital champions needed to generate innovative outcomes.

To be clear, the need for such digital champions is not limited to your IT group. They should exist throughout the broader organization. They need not be data scientists per se, but individuals with an innate understanding of the opportunities data and analytics create. The third layer of our Digital Transformation Stack are those digital applications that reside in various functional areas of the organization such as marketing, HR, finance, and operations. Each of these functional areas needs its own digital champions to recognize and execute opportunities for digital transformation. These digital champions should help build applications that advance decision making and ultimately create value.

Consider marketing. Digital marketing has become so ubiquitous that it is tempting to say that digital marketing is simply "marketing" these days. Digital marketers have a host of resources available to them. Point-of-sale data, such as scanners at the checkout counter, create voluminous data on individual purchasing behavior. Online sales channels are a vast treasure trove of user behavior. Companies are leveraging rich datasets on individual customers to customize advertising, create unique customer experiences, and recommend innovative new products and services. Sales teams are using predictive algorithms on whom to target and to generate leads.

Human resources professionals are finding new and creative ways to leverage vast internal datasets to improve employee productivity and well-being. They are leveraging data to improve workflows and assess individual performance. Amazon, for example, has sophisticated systems for tracking processes and driving efficiency in its warehouses. Uber leverages data across its user base to find efficient routes for drivers and to assess drive time. (In the next chapter, we will address some of the downsides for employees of such systems.) Knowledge management systems allow employees to have access to the vast expertise in the organization. Increasingly, these systems are leveraging AI to make automated recommendations to point employees towards useful knowledge.

In internal finance and procurement, digitization is creating new ways to improve financial processing. In retail, companies such as Square are simplifying the collection of over-the-counter payments while also improving the customer experience, all the while creating new data assets to feed into other efforts such as marketing and innovation. Salesforce and Workday provide cloud-based software-as-a-service solutions to manage your accounting needs and provide robust customer-managed relationships. Each of these tools promises to reduce bureaucracy, increase efficiency, and improve stakeholder satisfaction.

In operations and supply chain management, companies are using sensors and RFID technology to improve internal operations and to track goods as they make their way through the supply chain. Global shipping company Maersk partnered with IBM to create a blockchain solution to track goods as they move from manufacturers to distribution centers, to shipping ports, through customs, onto container vessels, and ultimately onto trucks to retailers or homes. The data being generated is revealing bottlenecks in the system and leading to innovations that improve efficiency. Predictive algorithms are helping to accurately forecast delivery times and to foresee potential supply chain snags due to extreme weather or other disruptions.

Regardless of the functional area, digital champions need to ask themselves some core questions about what they hope to achieve with a given digital application. Agrawal, Gans, and Goldfarb's Artificial Intelligence Canvas provides a useful framework for thinking through digital application projects (see Figure 5.2). To start, what is the task or decision that you are examining? What is a key uncertainty that could be resolved with more accurate prediction? What is the value of being right or wrong in this prediction? How will this prediction influence action? What outcome will be used to assess whether a digital application is achieving its desired results? With these questions framing the project, you can then turn your attention to the specific data needed. What data is available and needed to train your prediction algorithm? Once the algorithm is trained, what input data is needed to generate a specific prediction? What is the feedback mechanism to help improve the algorithm as new data is acquired, new predictions generated, and outcomes achieved?

Critical to any digital application is the role of judgment. You may recall from Chapter 1 that judgment is in the realm of humans and is complementary to any prediction generated. At my school, we conducted an analysis that found that the GMAT, the premier standardized test for MBA admissions, predicts only a small percentage of variance in career outcomes. The decision on whether to require the GMAT in our admissions process was a judgment requiring human decision making that was complementary to the analysis. The same is true of even automated decisions. The decision to automate is ultimately a human judgment. What are the costs of a poor automated decision? What risks and thresholds are tolerable for the decision being automated? These are for humans to answer, not machines.

One of the common challenges that companies face when pursuing a digital transformation is that all these efforts to wrangle data and create predictions is promptly ignored by organizational decision makers. A fancy data visual or robust prediction is useless if it does not have an impact on actual decisions. Potentially even worse are decision makers who blindly trust a prediction without questioning the data or basic assumptions behind it. Managing the human element—how we react to and use the predictions generated in a digital application—is a critical role for all digital champions.

What task/decision are you examining?

Briefly describe the task being analyzed.

Prediction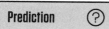

Identify the key uncertainty that you would like to resolve.

Judgment

Determine the payoffs to being right versus being wrong. Consider both false positives and false negatives.

Action

What are the actions that can be chosen?

Outcome

Choose the measure of performance that you want to use to judge whether you are achieving your outcomes.

Training

What data do you need on past inputs, actions, and outcomes in order to train your AI and generate better predictions?

Input

What data do you need to generate predictions once you have an AI algorithm trained?

Feedback

How can you use measure outcomes along with input data to generate improvements to your predictive algorithm?

What impact will this AI have on the overall workflow?

Explain here what impact the AI for this task/decision will have on related tasks in the overall workflow. Will it cause a staff replacement? Will it staff retraining or job redesign?

Figure 5.2. The Artificial Intelligence Canvas. Source: Based on Ajay Agrawal, Joshua Gans, and Avi Goldfarb, "A Simple Tool to Start Making Decisions with the Help of AI," *Harvard Business Review*, April 17, 2018, https://hbr.org/2018/04/a-simple-tool-to-start-making-decisions-with-the-help-of-ai.

THE DRIVE FOR GROWTH

So, you've built out your digital infrastructure, facilitating the collection and processing of your vast data resources. You've invested in data analytics and recruited digital champions who understand how to leverage data analytics to create valuable digital applications for the various functional managers of your business who eagerly embrace these new tools at their disposal. Yet all your efforts could be for naught. You could simply be throwing good money after bad competitive outcomes if you do not have a clear vision of what you hope to gain from digital transformation both in an absolute sense *and* relative to your competitors. For your Digital Transformation Stack to be complete, you must articulate your digital strategy and use it to guide your actions.

As discussed in Part I, your digital strategy should articulate a unique competitive position that is responsive to market opportunities and that allows you to appropriate some of the gains from your efforts. This digital strategy should drive your digital transformation efforts. All too often, I come across companies that seem to be doing digital transformation for digital transformation's sake. Often, they are doing so because they say they "have to" without thinking about how their offerings will be different or how their competitive position will improve as a result. Digital transformation is simply the latest hot trend requiring their organization's attention. This, in turn, can lead to technology driving solutions, rather than needs driving solutions. "We have a lot of data, we have data scientists, so let's create a whole bunch of cool applications and tools" goes the thinking.

Your digital transformation efforts need to have your digital strategy in mind. How is the market evolving? In what stage of the Competitive Life Cycle do you find yourself? What is the nature of competition likely to look like in a digitally transformed world? How are you trying to position yourself in this evolving market? What is your desired value proposition (your Why)? Who are your desired customers (your Whom)? What is your business model for monetizing your offerings (your What)? How will you capture value from your efforts, generally, and what capabilities do you need to develop specifically (your How)?

Collectively, the answers to these questions should help you prioritize which digital applications to develop. Will building this new application help us deliver value to customers in some new way consistent with our overall value proposition? Will building this new application help drive efficiencies in the organization and outcompete rivals? Will building this new application allow us to capture greater value from our efforts, more generally? What is the opportunity cost of building this application versus alternative applications that we could focus on? Ultimately, does this application best drive top-line and/or bottom-line growth for the organization?

Consider a new digital application that is created for your HR team that allows employees to easily request vacation days and track their time in and out of the office. The application can be used to make accurate predictions on daily staffing levels. The hope is that the system will help avoid bottlenecks, create staffing efficiencies for managers, and increase employee satisfaction by reducing the administrative burdens on them. And let's assume that the application delivers on this premise. All good, right? But while your digital transformation team is creating office efficiencies, a competitor has just reenvisioned their business model, leveraging consumer data, to build an online application that automates work processes and upends the need for heavy staffing. Suddenly, your efforts look like rearranging the deck chairs on the *Titanic*. Making the ship's engines run more efficiently isn't very helpful if the ship is heading towards an iceberg.

In a world of constrained time and resources, where you place your digital transformation efforts is critical. How do you avoid the iceberg? Your digital strategy should be your guidepost. It should help you prioritize which projects to invest in and which projects to forego. How will a specific project advance your digital strategy? How do you select a portfolio of digital transformation projects that lead to digital applications that help you achieve a valuable competitive position, creating and capturing value and ultimately driving growth?

Not to say that your digital strategy is a static target. As we've discussed in previous chapters, the digital age is marked by rapid upheaval and the need for continually innovating to meet ever-evolving market conditions and opportunities. Your digital strategy should serve as your guidepost, but you must be prepared to update your strategy as technologies advance and competitors act and as you learn better the potential and limits of your own organization. You must be willing and able to adjust your strategy as conditions evolve. In the digital age, such agility is paramount.

THE QUEST FOR AGILITY

For the modern manager, the pursuit of agility has become an epic quest. Silicon Valley upstarts talk of failing fast and pivoting quickly. Innovate or die is their mantra. As discussed in Chapter 4, innovation has evolved from a narrow focus on corporate R&D to a broader focus on open innovation in which new ideas are sourced from a broader innovation ecosystem.

Firms are trying to balance both exploitation and exploration, balancing portfolios of innovative projects that help them advance today and prepare for tomorrow. They are looking to build what Michael Tushman and Charles O'Reilly call organizational ambidexterity—the "ability to be efficient in . . . management of today's business and also adaptable for coping with tomorrow's changing demand."[7]

Numerous approaches have been proposed to help build such ambidexterity. Design thinking is one of the most prominent, popularized by companies such as IDEO and organizations such as Stanford University's d.School. My colleague Jeanne Liedtka has been one of the pioneers in applying design thinking principles to business innovation. Design thinking is a process for solving problems by focusing on user needs and relies on observing, with empathy, how people interact with their environments.[8] Companies and organizations use this approach to innovate new products and services, develop digital applications, and, more generally, refine their overall strategy. I frequently use design thinking concepts in working with organizations in strategic planning. See Figure 5.3 for a summary of the standard design thinking approach.

The design thinking process begins with empathetic research, observing and engaging with people to understand their experiences and motivations. Ethnographic approaches in which you observe people as they engage with their environment are common, for example, observing customers as they use your products in the real world. Interviews and surveys are also common approaches, as is collecting archival data, perhaps from your own data lake. The goal is to identify pain points for the user. What challenges do they face? How do current offerings meet their needs? How do they fall short?

With a better understanding of these challenges, the next step is to define design criteria that specify the attributes of a favorable solution to address users' challenges. The key is to create criteria that specify outcomes and not solutions. For example, upon observing that supply partners express frustration with invoicing and payments, you do not immediately jump to designing an automated solution but unearth the specific criteria by which your partners would judge a successful process—perhaps reducing time spent on those activities.

With design criteria in hand, you can then turn your attention to ideating solutions and concepts. There are several approaches to ideating, such

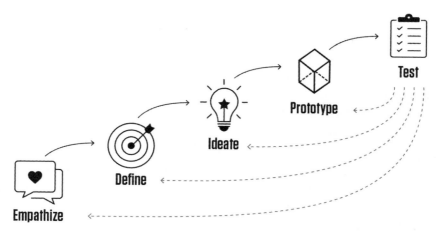

Figure 5.3. The Design Thinking Approach.

as brainstorming and mind mapping. Generally, the key is to start with a broad open-ended process to help stimulate free thinking and to drive "outside the box" thinking, and then to winnow the ideas generated to a manageable set of concepts for further explication. Napkin pitches are one way to more fully flesh out a specific concept specifying the target user and their unmet need, how the concept will create value for them, how you plan to execute the concept, and whether if properly executed the concept will provide a differentiable advantage over competitive offerings. Central to concept development is to identify the underlying assumptions associated with the concept, that is, what needs to be true for the concept to be successful.

With well-specified concepts in hand, it is time to prototype and test them. Primacy is placed on cheap and quick experiments that test the underlying assumptions identified. All too often, organizations opt for expensive pilots or, even worse, immediately jump to implementation. In the digital age especially, there is a tendency to immediately develop a full-functioning application. This is typically unnecessary. Wireframes—a simple two-dimensional illustration of an application—can be sufficient to garner user feedback and test assumptions about value creation. Further research such as user surveys or competitive benchmarking may be sufficient to test assumptions and determine whether further development of testing of a concept is warranted.

Last, but certainly not least, is iteration. At each phase of the design thinking process, one may wish to revisit a previous stage, conducting further empathetic research, refining design criteria, ideating or developing new concepts, creating new prototypes, and executing new tests. The key is to emphasize rapid learning. If a given concept is proving less promising, be quick to retire the concept and move onto a new one. Even better, use what you've learned from a failed concept to develop a better new concept.

Central to design thinking is the creation of a learning culture. Even failed projects should be viewed as a positive if they help advance learning and identify unfruitful paths to avoid in the future. One key is to monitor projects and be prepared to kill them quickly if needed; in other words, to fail fast. Another key is to celebrate learning, rewarding those who enhanced learning even if a given project fails. This approach can be very hard for many organizations to embrace. A prominent science-based company that I have engaged proudly talked about how they had changed policies and now everyone was allowed one failure—but only one. Not exactly the spirit of celebrating learning intended! Agile organizations should encourage a climate of risk taking, to encourage the exploration of new concepts and not to fear failure.

One way to create a learning culture is to assign individuals to more than one project team. In that way, failure is not catastrophic but simply frees up more time to work on other, more promising projects. This can incentivize teams to learn quickly and rapidly kill unpromising projects. Another strategy for creating a learning culture is to have a product orientation and not a project orientation. In other words, focus on outcomes and not simply the completion of the project. Cross-functional teams that span units of the organization may be helpful. A digital transformation project that spans, say, marketing, operations, and logistics may be more likely to be judged by its overall contribution to the organization's strategy than by merely checking a completed task box for a given unit.

Overall, the quest for agility must operate at multiple levels, from the C-suite down to the employee on the shop floor. To be clear, digital transformation is not limited to the IT organization or a data analytics group. It is the responsibility of the entire organization. Leading a digital transformation is as much about how you create the conditions and culture in which everyone in the organization understands and is supportive of what you are trying to achieve. It is about creating an agile organization focused on learning that is willing and able to pivot as roadblocks are encountered and opportunities arise. Digital transformation needs digital leaders.

THE HUMBLE LEADER

So, what attributes define digital leaders? Ed Hess and Katherine Ludwig, in their book *Humility Is the New Smart: Rethinking Human Excellence in the Smart Machine Age*, argue that humility is central to successful leaders in the digital age. By humility, they do not mean meekness or lacking confidence as all too often is assumed in popular culture, but rather a mindset that is open, is self-accurate, and enables one to embrace the world as it is.[9] The digital age moves fast. Understanding that you do not have all the answers or know all the facts is critical. Overconfidence can prove fatal. You may be smart, but there are always others who know more in various domains.

Hess and Ludwig identify four "new smart" behaviors that define the humble leader: quieting ego, managing self, listening reflectively, and embracing otherness. Quieting ego is about how one can deliberately reduce emotional defensiveness and be more empathetic and open-minded. Managing self refers to the ability to self-regulate one's emotions, thoughts, and behaviors, in many cases to slow-down, not jump to conclusions, and think deliberately and deeply. Listening reflectively is not just about listening to others, but truly caring about what the other person has to say and to listen in a nonjudgmental way, to truly learn from others. Last, embracing otherness is about valuing diversity of thought and experience, building trust, and conveying caring.

Hess and Ludwig emphasize that each of these behaviors can be learned. Digital leaders are not just born, they can be made. Like a world-class athlete, humble leaders practice their craft, working to deepen and improve their behaviors. They encourage others to embrace humility and develop their own behaviors. They focus on building connections, high-value partnerships both within and beyond the organization. The humble leaders know that human development capabilities are more likely to be a source of competitive advantage than technological expertise. The best organizations will be those that can "attract, develop, and retain the best human learners, thinkers, and collaborators."[10]

As more and more tasks are assigned to smart machines, Hess and

Ludwig see a decided turn towards the human side of organizational excellence. In the same way that prediction and judgment are complements, organizations will have to embrace thinking, creativity, and innovation as complements to the powerful computational machines that we are building. They identify a set of organizational capabilities as critical "smart machine age" skills: higher-order critical thinking, creativity and innovation, and high emotional engagement with others. Together, these are what will help an organization be able to learn, adapt, and innovate to meet evolving stakeholder needs faster than competitors.

FRAMEWORK 5: PLANNING YOUR DIGITAL TRANSFORMATION

At this point, you have articulated a desired competitive position in a digitally transforming world specifying your unique value proposition, your target audience, your underlying business model, and the basis for your competitive advantage (see Framework 3). You have identified a list of offerings and capabilities required to establish this position and have mapped out a strategy for developing these offerings and capabilities (see Framework 4). Next, we get more granular, identifying specific digital applications that we need to develop in support of these offerings and capabilities (see Figure 5.4). For example, for our new venture in autonomous luxury limousine services, perhaps we need to develop a mobile application to deliver a superior service experience that delivers on our brand promise and differentiates us from our competitors.

To begin, you need to identify a proposed suite of digital applications required to support your overall digital strategy. To generate this list, think back to design thinking principles. Develop design criteria by which you would judge potential applications. Engage in ideation exercises, such as brainstorming, to generate a plausible list of applications. Refine the list by engaging in combinatorial play between concepts, generating new ideas, and eventually selecting a finite set of ideas for proposed applications worth further exploring. Enter these applications in column one along with a brief description of each concept.

Next develop the "napkin pitch" for each application by explicitly identifying the assumptions undergirding the concept in the boxes in columns two through five. Consider four broad categories of assumptions. Data

Application	Data Assumptions	Value Assumptions	Execution Assumptions	Defensibility Assumptions
Describe a specific digital application that may be created for the enterprise.	What data is required to power the application? Do you need training data to develop an algorithm? Does the application capture new data to improve its performance?	How will this new application deliver value and to whom, e.g., customers, employees? Will this new application drive efficiencies in the organization?	What must be true for us to execute this application? What people, processes, and systems are required of the organization?	Will this application provide a competitive advantage and, if so, how? Will it help solidify our aspired competitive position in the market?
See above.	See above.	See above.	See above.	See above.
See above.	See above.	See above.	See above.	See above.
See above.	See above.	See above.	See above.	See above.

Figure 5.4. Planning Your Digital Transformation Framework.

assumptions speak to the specific data needs of a considered digital application. Value assumptions are those that specify how the application will deliver value to a specific stakeholder or stakeholders. Execution assumptions refer to what must be true to be able to execute the application, such as what people, processes, and systems are required of the organization. Defensibility assumptions refer to how this application may provide a competitive advantage and may help you achieve your desired competitive position in the market. In all cases, think about your Digital Transformation Stack. What is required of your underlying digital infrastructure? What demands do your data assumptions make on your data lake and cloud services? How about your data analytic capabilities? Do execution assumptions rely on building out a specific capability? How about digital champions? What is needed to make this application a success?

With your assumptions articulated, consider ways that you can quickly and inexpensively test these assumptions. What can you learn before building a full prototype? Are there ways to test your value assumptions by sharing rudimentary descriptions with customers, such as wireframes in the case of software applications? What internal analyses can be performed to de-risk your execution assumptions? What competitive intelligence is necessary to test defensibility assumptions? For each set of assumptions for a given proposed application, identify the tests you plan to perform to verify or refute an assumption and place them in the boxes provided. The key is to privilege learning. What is the easiest way to test the assumption and provide valuable feedback in the viability of a proposed application? In all cases, work iteratively. As you begin to test assumptions and learn more, revisit your proposed applications. Do they need to be adjusted or even abandoned? In the spirit of Silicon Valley, fail fast and move on.

6

Policy in the Digital Age

There was a time when Facebook was going to save the world. In January of 2011, a group of dissidents in Egypt began using Facebook to organize protests against the current government and its leadership. Their efforts soon ballooned into a massive protest movement spreading across the Middle East. All the while, relatively new social media applications such as Facebook and Twitter had become critical to organizing protests, building support within countries, and communicating with the broader world. With each post, tweet, and YouTube video highlighting the brutal response of the Egyptian government to the protests, global sentiment turned in favor of the protesters. On February 11, the president of Egypt, Hosni Mubarak, resigned.

In the immediate aftermath of the Arab Spring, *Time* magazine declared "the Protestor" its 2011 person of the year. Facebook and Twitter were cited as transformational technologies that would return power to the people, help encourage governmental transparency, and usher in a new era of freedom and liberty. Some even referred to the events of that spring as the "Facebook revolution." Silicon Valley and its culture of innovation were providing the tools and technologies to transform society for the better. Digital companies were not just building the world's next great products and solutions, they were solving the world's most vexing problems. The digital age was going to be our savior.

Just as the Arab Spring failed to deliver vibrant free societies, digital technology soon found itself the subject of scorn, not celebration. Facebook's moment in the sun quickly faded. Researchers were finding evidence that social media was creating unhealthy social comparisons, especially among teens and young adults. Some accused Facebook of enabling self-harm and even suicide. Users were finding Facebook to be a forum for the most outrageous of claims and the most vitriolic of comments. CEO and founder Mark Zuckerberg was hauled in front of the US Congress to explain Facebook's algorithms for prioritizing content and to explain how it planned to remove untruthful and harmful posts. Criticism did not end there. Data privacy, employee health and wellness, use of facial recognition technology, tax avoidance, participation in governance surveillance programs, and censorship policy, among others, had become flashpoints for the company.

Facebook was not alone. All the large technology companies found themselves under greater scrutiny. Google for antitrust and data privacy. Apple for exploiting partners in its App Store. Amazon for its labor practices in both its fulfillment centers and corporate offices. Uber for a toxic work environment that fostered sexual harassment. DoorDash for its refusal to treat drivers as employees. WeWorks for overhyping its business results. Airbnb for potential risks to clients from unscrupulous and even predatory renters. The list goes on and on. For each of the companies cited above, I could cite a dozen other complaints levied on them. Each of these challenges highlight that the digital age is not without peril. Some of these challenges reflect bad management practice and unethical behavior found unfortunately in any age. Some, however, reflect specific issues that are, at the very least, more acute in the digital age. Any company operating in the digital age must understand, account for, and act upon these issues if it is going to flourish.

Complaints are coming from a diverse set of stakeholders. Customers concerned about their personal and digital security. Employees and partners worried about being exploited. Activists looking to advance their specific causes by highlighting the behavior of individual organizations. A news media eager to expose and amplify any concerns. Elected officials responding to their constituencies and, perhaps on occasion, looking to make a name for themselves in the broader public arena. Regulators and other public officials trying to make sense of the applicability of existing rules to the digital arena. Even investors who are growing concerned with the environmental and social performance of companies in which they invest.

For the manager, each of these stakeholders presents risks and opportunities, as important if not more important as what is typically viewed as normal business risk. Like all risks, these "nonmarket" risks need to be understood and managed. We say "nonmarket" to highlight that they are often thought of arising in arenas outside the daily operation of business, such as risks arising from activists or policy. However, the term is misleading because these risks are simply the norm when conducting business in the digital age. Your digital strategy would be incomplete if it did not account for and address these nonmarket forces. Thoughtful CEOs know this and actively manage their broader "nonmarket" strategy.

THE END OF PRIVACY

Among the many perils of digitization, perhaps most important is what some are calling the end of privacy. Central to most of the strategies highlighted in this book is the acquisition and analysis of data that allows for the creation of value-added services. On the surface, this need not be problematic. A company using sensors to monitor manufacturing equipment to identify maintenance issues and to help improve production efficiency is unlikely to prove controversial. However, when data comes from individual users and is used to reveal individual preferences and behavior, issues of data privacy become paramount.

What are, and are not, proper applications of user data is a hotly debated topic. On one side, user data can help with the customization of products and services to a user's specific wants and desires. Think of Spotify's custom playlists based on your demonstrated listening preferences. To many, even a targeted ad for a new dress on Instagram is not an inconvenience but a delight. On the other side, of course, are the many ways that user data can be misused. One widely spread story tells of a father who was furious at Target for sending pregnancy ads to his teenage daughter, until it was revealed that the store's algorithm had accurately predicted her pregnancy.[1] More insidious stories abound of employees listening to and sharing private recordings from smart speaker systems.[2]

For companies such as Google and Facebook, user data is central to their business model of selling targeted advertisements to run on otherwise free services. Efforts to give users the ability to control what data is collected and stored by these companies can have a direct impact on the efficacy of the advertisements they provide, lowering the attractiveness of these platforms to advertisers and ultimately affecting the price and demand for advertising on these sites. This has given rise to what Harvard Business School professor Shoshana Zuboff refers to as "surveillance capitalism," which she defines as "the unilateral claiming of private human experience as free raw material for translation into behavioral data. These data are then computed and packaged as prediction products and sold into behavioral futures markets—business customers with a commercial interest in knowing what we will do now, soon, and later. An insatiable appetite for user data to make better predictions of their behavior."[3]

Zuboff highlights that surveillance capitalism is not only about making passive predictions on user behavior, but increasingly about finding ways to shape and direct user behavior. The big tech companies have hired scores of psychologists and behavior economists to experiment with "nudges"— subtle interventions desired to shape your behavior. While sometimes those nudges can be towards arguably desirable ends—reminding you to save money for retirement or motivating you to work out—they can also be used to influence what you buy and shape your preferences. This was never more evident than in the 2016 US presidential election, when it was discovered that foreign actors were purposefully sowing discord on online platforms such as Facebook and Twitter and even trying to incite citizens to take to the streets and engage in violent protest.

Surveillance capitalism is not an unavoidable outcome of digital technologies. It is a choice by businesses. For a business, it derives from the choice of a business model predicated on exploiting user data. Often policymakers are complicit, allowing businesses to pursue such strategies and failing to provide safeguards that allow users to truly own their data. It doesn't have to be that way. Consider Apple. Apple's business model is largely to sell hardware devices to customers. This helps mitigate some of the demand for exploitation of user data. Apple can use your data to improve the products it provides to you while also promising you a certain degree of protection for your data, such as shielding it from other parties. It can empower data owners. This tension between business models came to a head in April 2021 when Apple allowed its users to easily opt out of data tracking by applications as you use your mobile device. Over

96 percent of Apple iPhone users opted out of such tracking. This has had a large material impact on companies such as Facebook, which forecasted a $10 billion decline in 2022 advertising revenue because of the change in Apple's data policy.

Of course, even when a company promises to keep your data private, there is the question of how protected your data is from cybersecurity threats. In 2013, Target was hit with a massive cyberattack that compromised forty million customers' credit and debit card numbers. In an analysis of what happened, it was determined the malware that perpetrated the attack entered Target's systems through a personal computer attached to the HVAC system at one of its facilities by an unaware air conditioner repairman. Within hours, the malware had moved through the HVAC system and into the company's servers, ultimately infiltrating the cash registers in its network of stores.[4]

A massive shadow industry of illicit hacking now spans the globe. Supported by the dark web and cryptocurrency, the hacking market has evolved into a sophisticated international exchange. Some estimates put the money collected worldwide from ransomware at $18 billion,[5] nearly $350 million in the United States alone.[6] Global crime syndicates have been found to deploy sophisticated cyberattacks against companies and governments around the world. Some of these hacker groups, with names like DarkSide and REvil, are selling their services to the highest bidders.[7]

Of course, the most severe privacy risks may still come from within the organization. Sensitive health information could be misused by employers. Facial recognition could be used to track the movement of customers and employees without their knowledge or consent. These concerns have become particularly pronounced as these technologies have moved into the public arena, adopted by governments and police departments. In the United States and abroad, we are seeing the increased use of facial recognition and other data-driven techniques to try to identify and track criminals. "When does the right to privacy begin and when does it end?" is a question all organizations will have to grapple with in the digital age.

'THE TERMINATOR' AND OTHER DYSTOPIAN FUTURES

The end of privacy and the misuse of data are central to many dystopian futures portrayed in science fiction. George Orwell's *1984*, published in

1949, foresaw a future in which an authoritarian government used omnipresent surveillance and manipulation of facts to achieve near total control of society. Given facial recognition software, the ubiquity of cameras in public, and the building of massive datasets, such a surveillance state may not seem so far-fetched in our digital age.[8] In the United States, police are using facial recognition software to identify, track down, and arrest suspected criminals. In the United Kingdom, closed-circuit cameras have been used for decades to monitor public spaces. China has gone further than almost any other state to track and monitor its citizens, creating a "social credit" score with which demerits can be assessed for acts such as jaywalking. They are even creating analytics to predict when people might protest or break the law *before it happens.*[9] Orwell's creations such as the "thought police" may feel a little too on-point in a world dominated by social media where the truth is constantly contested and restructured.

Some technologists worry about a fate even more grave than totalitarianism: the technological singularity. Popularized by mathematician John von Neumann, the singularity refers to a self-enhancing artificial intelligence that quickly cycles through improvement cycles and eventually surpasses human intelligence, far outstripping our problem-solving capabilities. Some, such as Elon Musk and Stephen Hawking, have expressed the view that the singularity could lead to the extinction of humans, with Musk musing that we may already be living in some Matrix-like simulation.[10] The singularity is seen as a possible solution to physicist Enrico Fermi's famous paradox—if the universe is so vast, why haven't we discovered other intelligent life? In popular culture, this is often portrayed as a superintelligent AI deciding to end human existence—think of Skynet in the *Terminator* movie franchise. Perhaps this is the fate of all intelligent life.

Fortunately, many leading thinkers on artificial intelligence believe that such a scenario is far-fetched. AI systems today are what are referred to as narrow, or weak, systems. They are designed to solve problems assigned to them. General, or strong, AI systems that display something like human consciousness have proven elusive. Computer scientist Stuart Russell in his book *Human Compatible* points out that an AI gaining consciousness and deciding to end human existence is highly unlikely.[11] Much more likely, an AI will do exactly what it is programmed to do by humans. Unfortunately, this should also give us pause. Take for example, Netflix's *Black Mirror* television series episode titled "Metalhead." In the episode, vicious robot dogs terrorize the countryside, hunting and killing any humans they find. In this scenario, the robot dogs are simply executing the instructions given to them, presumably by an army deploying them against an enemy on the battlefield. Implied is an inability to halt their preprogrammed behavior if they escape from the arena.

While such dystopian futures may seem exclusively the purview of science fiction, our ability to control and understand the AI systems we create is decidedly not. When and how to use autonomy in warfare is an active discussion within the military. Autonomous vehicles must ultimately make decisions that have an impact on life and death. The trolley problem is a classic moral dilemma, in which the pilot of a runaway trolley car must decide whether to plow into a crowd of people on the street or divert the trolley into a brick wall killing all those onboard. There are many manifestations of the trolley problem that vary the number of people involved or their background (for example, family members versus strangers). While there is no obvious "correct" answer to the trolley problem, what is clear is that an autonomous vehicle will need to have a solution to the trolley problem hard coded into its software.

The ethics of algorithms is a growing area of study. There is evidence that improperly trained facial recognition systems may lead to more false positives for certain groups of people, demonstrating racial and gender bias. This is deeply concerning, as such systems increasingly are being used in law enforcement. Compounding these concerns are that a lot of approaches to artificial intelligence, such as machine learning, are "black-box" techniques. In other words, the relationship between cause and effect is not specified, but rather emerges from the data. As a result, such systems end up generating correlations between inputs and outputs without any analysis as to whether they are causal. Say, for example, a credit card company's algorithm predicts fraud on the basis of the color of the cardholder's eyes. Not only is such a relationship very unlikely to be causal, the company may not even know that their algorithm is using such a logic since it is arising out of massive data feed into training the system. Biases may emerge without any awareness by the organization deploying the AI.

Even when our digital systems work as planned and do not demonstrate biases, there are concerns that they may work too well. There are growing concerns that AI and related technologies will lead to a jobless future as more and more jobs are automated or eliminated. Concerns about the displacement of jobs by technology go back to at least the Industrial Revolution. The original Luddites were English workers who destroyed cotton and woolen mills that were threatening their jobs in the early 1800s. History has proven repeatedly, however, that new technologies that reduce labor needs often ultimately create more jobs as they help generate economic growth and new opportunities. Fear of technology replacing jobs, at least in the aggregate, have largely been misplaced over the years.

Yet some are arguing it is different this time. The scale and pace of change are unprecedented. The global industrial robot market is projected to more than double in size over the next six years, from $40 billion to more than $100 billion.[12] Routine knowledge work such as accounting and paralegal services finds itself being automated. Think of the impact of TurboTax on the preparation of tax returns by accountants. In 2011, an influential study suggested that driving a truck was largely automation-proof.[13] Ten years later and people write of the eminent decline in truck-driving jobs as automation takes hold. Fear of a jobless future has been gaining currency in Silicon Valley and beyond. The call for a universal basic income from 2020 US presidential candidate Andrew Yang is frequently echoed by technologists and entrepreneurs concerned about digital displacement and a future in which large numbers of people are unemployed due to technology.

THE RISE OF THE REGULATOR

As concerns about the societal cost of digital disruption rise, so do calls for public action. Antitrust regulators in the European Union and the United States are turning their attention to the technology sector. In the simplest terms, antitrust is concerned with two things. The first is the consolidation of significant market power. The second is exploiting that market power or, as I like to say, "behaving badly." The first is undeniable among technology firms. As discussed early in the book, the presence of network externalities can lead to winner-take-all dynamics that leave technology companies with dominant market shares. The second is more contested. Amazon, for example, claims that its significant position in online retail leads to lower prices for consumers due to scale and efficiency. The consumer is benefiting, not being exploited, by Amazon's strong position, it argues.

There is increasing awareness that antitrust laws designed for the robber baron era are ill-suited for the digital age. The concern a hundred years ago was with companies building up huge-scale economies by consolidating the competition and charging monopoly prices to downstream buyers. Think John D. Rockefeller's Standard Oil Company or Andrew Carnegie's Carnegie Steel Company. In the digital age, it is not clear if monopoly pricing is occurring or if the consumer is being harmed. In the case of Google and Facebook, it is not even clear who is the consumer—the user or the advertiser? What if it's the case that those exploited are suppliers or partners, and downstream users are rewarded with superior products and services at low prices?

In the digital age, scale advantages are largely based on data rather than physical assets. The network externalities from data are a natural by-product of the technology and hard to eliminate. We have seen this in the telecommunication sector, in which AT&T recobbled itself back together after being forced to break up into eight separate regional companies in 1984. One could imagine a similar result if, for example, Facebook was forced to spin out Instagram or WhatsApp. Efforts to constrain technology companies may be thwarted, as the advantages to consumers of having their data collected at scale to feed predictive algorithms to create value-added services constantly drives scaling. Antitrust becomes something like a game of whack-a-mole in which a new scaled technology company pops up after every effort to bring down an existing technology giant.

Though such entrepreneurial replacement is looking increasingly less likely. Big tech companies often gobble up smaller ventures before they reach a scale to truly threaten the existing players. In the last decade, we have seen massive acquisitions by existing tech giants. In 2014, Facebook acquired WhatsApp for $22 billion. In 2017, Intel bought AI pioneer MobileEye for $15 billion. In 2019, Microsoft acquired GitHub for $7.5 billion. In 2020, Salesforce acquired Slack for over $27 billion.[14] And there assuredly will be many more. This has helped promote stability in the ranks of big tech. By one estimate, since 2000 the "chance that a high-ranking firm (measured by sales) will drop out of one of the top four spots within four years has fallen from over 20% to around 10%."[15]

Even more distressing is the potential for horizontal diversification, in other words, technology companies diversifying into adjacent industries. While antitrust is largely concerned with the exertion of market power in a specific market, horizontal diversification is about exploiting market power across markets. With data as the engine for growth, digital companies are realizing that they can exploit their vast reserves of data to enter other industries. Simply put, the more they know about you as a customer, the more value-added services and products they can design for you. Hence technology companies such as Apple, Amazon, and Google have entered media and entertainment, financial services, education, health care, and transportation, among many others. This is raising concerns that a handful of technology companies are going to dominate the global economy.

Recognizing the growing—and somewhat hard-to-contain—power of these technology companies has led to increased calls for direct regulation. Every day, new rules and regulations are arising with regard to the use of private data. The European Union's General Data Protection Regulation went into effect in May 2018 and governs how personal data of individuals may be processed and transferred. The law requires, among other things, the consent of subjects for data processing, the anonymizing of collected data to protect privacy, notifications when data is breached, and the appointment of a data protection officer for some companies.[16] In the United States, the California Privacy Rights Act passed in 2020, giving individuals the right to rectification, that is, a consumer's right to correct inaccurate personal information, and the right to restriction, that is, a consumer's right to limit the use and disclosure of sensitive personal information.[17] While there is no single federal US law regulating data usage, pressure grows to harmonize standards within the United States and beyond.

On March 24, 2022, the European Union released the final text on what could be one of the most transformational regulations, the Digital Markets Act (DMA). The DMA creates a set of rules for "gatekeepers" operating in the EU, platforms with greater than $7.4 billion euros in annual revenue and at least forty-five million monthly users.[18] Specifically, gatekeepers will be restricted in their ability to share data across platforms and from requiring businesses using their platform to use their payment services (like Apple's App Store). Furthermore, they need to give advertisers and publishers access to some of their algorithms. The DMA is expected to be implemented in 2023.

For social media companies, there are increasing regulatory pressures around curating the postings on their platforms to flag or censor harmful or untruthful content. To date, most of the pressure brought by government authorities and independent activists have been for companies to self-regulate their platforms. However, pressure is mounting in various jurisdictions to pass legislation to set standards for social media platforms. In the United States, there remain several unresolved questions, such as which federal agency would oversee social media companies. Are they telecommunications companies, in which case the Federal Communications Commission would have dominion? Should they be regulated by the Federal Trade Commission, which regulates a vast number of businesses and industries?[19]

For two-sided market makers such as Uber and Airbnb, there are concerns about whether supply partners are truly independent or constitute employees and formal partners. For Uber and other ride-hailing and food-delivery service providers, there have been several recent legislative attempts to force companies to treat drivers as employees. The most prominent effort was Proposition 22 in California in 2020, where voters ultimately chose to allow companies to treat drivers as independent workers. However, the issue appears to remain far from settled. For Airbnb and others, there are questions as to whether they are legally responsible for criminal behavior by their supply partners. Many US states and local municipalities have passed legislation to create certification processes for hosts and to clarify liability questions.

Each of these examples highlights an inherent tension with digital platforms: How responsible are platform owners for the behavior of users on their platforms? Many technology companies would prefer to be simple "connecters" bringing users together or buyers and sellers. If a transaction or exchange or posting is harmful or disagreeable, that is between the offended and offending parties. The platforms, themselves, bear no responsibility. Increasingly, such a stance is viewed as untenable, even immoral. What is clear is that political and societal actors are increasingly litigating these issues and pressuring platform providers to act. See Figure 6.1 for a summary of the various challenges posed by digital technology.

THE NONMARKET STRATEGY TOOLKIT

In such a brave new world, managers need to be deliberate in how they think about the myriad of stakeholders making demands of their organization. A helpful place to start is with a classic strategy framework, a stakeholder

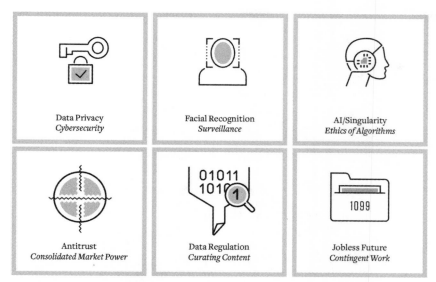

Figure 6.1. Digital Age Challenges.

analysis (see Figure 6.2). A stakeholder analysis identifies your primary and secondary stakeholders and analyzes their motivations: What do they care about? What do they value? Primary stakeholders include those with whom you have a direct transactional relationship, for example, suppliers, employees, and customers. Secondary stakeholders are those who have an impact on and are affected by your actions either directly or indirectly and include activists, regulators, the media, trade associations, and so on.

Employees, for example, likely care are about their pay, their job security, their opportunities for advancement, and the impact the company has on the broader world. Customers care about the price paid and the value delivered for goods and services. They may also care about how their purchases reflect their broader sense of identify and values, what consumer behavioralists refer to as value-expressiveness or ego-expressiveness. Thus they may avoid "good" products from "bad" companies. Activists care about their motivating issue, perhaps personal privacy, and may desire to raise awareness or motivate legislative action even if their actions are targeted at an individual firm. Politicians may be balancing several concerns including how to appease donors and how to get reelected.

With your stakeholders assessed, you can begin to formulate responses to the various demands they may place on the organization. See Figure 6.3

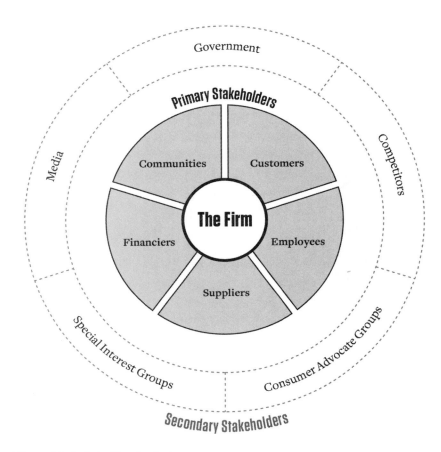

Figure 6.2. Stakeholder Analysis.

for a summary of some common stakeholder strategies. A classic typology of responses is fight, flight, or accommodate. To fight, as the name implies, is to resist stakeholder demands. In the political arena, this may entail engaging in lobbying for desired legislative outcomes. In the case of California Proposition 22, a coalition of gig companies spent $225 million to fight for the classification of drivers as independent contractors.[20] Flight refers to strategies to avoid confrontation all together. In some cases, this may mean dropping a controversial product or exiting a challenging industry. The chief technologist for business software maker LivePerson admits to dropping several AI products because of privacy worries.[21] Last, to accommodate is to try to meet some or all demands of various stakeholders. For employees, that may entail paying higher wages. For users,

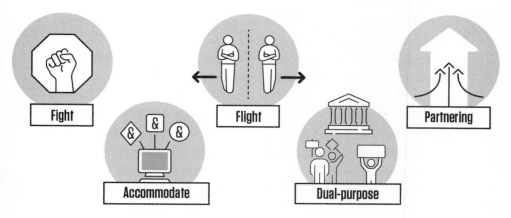

Figure 6.3. Nonmarket Strategies.

that may entail giving them control of their private data. For regulators, that may mean working to help devise smart legislation.

Accommodate strategies should not be underestimated. My colleagues Ed Freeman and Bobby Parmer, in their book *The Power of And*, highlight pursuing opportunities that create shared value.[22] Rather than view the world as a zero-sum game, can you envision a new outcome that jointly creates value for stakeholders? Perhaps you can innovate in ways that remove sources of potential conflict? Your competitive position and business model may make it more natural for you to accommodate certain stakeholder demands. Apple's emphasis on selling high-end devices makes it a natural partner for those looking to ensure data privacy in a way that Google is not.

Often, partnering can be central to an accommodate strategy, even partnering with competitors. Working together, you can help shape the institutional envelope in ways that level the playing field on some dimensions and create new opportunities for competitive success. In the digital world, debates around net neutrality are illustrative. For content providers including many digital platforms, net neutrality removes the "pipes" to your home and devices as a basis of competition, removing a potential bottleneck and allowing companies to compete online (rather than fighting for access to your home). Of course, there are vested interests on the side of infrastructure providers, such as cable companies, to resist the push for net neutrality. Coalitions of competing interest groups are common in the digital age. Accommodating some stakeholders may imply fighting others.

This highlights a fourth strategy beyond fight, flight, and accommodate—what economist David Baron refers to as a "dual purpose" strategy.[23] A dual-purpose strategy is one that provides benefits in both the market and institutional environments. Using Uber as an example, Baron highlights how it used fast entry into various local markets to build stakeholder affinity, specifically with customers who became fans of the service and willing advocates for their presence in the local market. When resistance from other stakeholders invariably presented itself, for example, when local taxicab associations would try to ban Uber from operating, Uber would lobby and fight restrictions by mobilizing its customer base to advocate on its behalf. Airbnb, Amazon, and others have used similar strategies.

While such dual-purpose strategies are nothing new, they can easily be criticized as ways of capturing or buying off stakeholders. Uber has faced withering criticism for its aggressive stakeholder-engagement tactics. This highlights what might be called the "dark side" of nonmarket strategy. The very tactics that may help you flourish in the digital age, while pleasing to certain stakeholders, can also run afoul of others. What is "fair" and "unfair" is often in the eye of the beholder. Even simple logics such as "obey the law" can be suspect in a world where organizations routinely actively work to shape the law to their liking.

THE VALUE OF VALUES

What is a responsible leader to do? There is no simple answer. Your values as an individual and organization ultimately guide what actions you feel are appropriate. This is the reason that we place values at the top of the Strategist's Challenge. They are your north star. Your guiding light. They determine which competitive positions you feel comfortable pursuing and which you do not. Which competitive actions to take and which to avoid. While your strategy may constantly adjust to meet evolving market needs, your values should be eternal. If your values do not align with a given market opportunity, maybe it is time to look for a new segment in which to compete.

Too often, managers fall back on bromides such as "doing [whatever] maximizes shareholder value." To be clear, maximizing shareholder value is not a legal restriction about what is required of you as a manager. Managers are required to be fiduciaries to their shareholders. Technically that means they are required to be trustworthy and transparent in their actions. Yes, shareholders can direct their capital elsewhere or even remove leaders with whom they disagree, but that does not permit the manager to avoid value-based decisions. Even reasonable guidance such as "maximize profits while not engaging in illegal activities" gets murky when a company can actively influence and shape the legal environment it operates in through lobbying and other forms of persuasion.

Thus the pursuit of profit does not absolve the manager of making moral decisions about what activities are consistent with their values and which are not. Many well-intentioned managers have found themselves going down the slippery slope of moral relativism: "Our customers love our new service based on their personal data, I know it is a little unseemly, but what if we grabbed a little more personal data so we can improve the service even more!" Just because you can leverage data to provide a new product or service, does not mean you should.

The digital age presents many moral challenges to leaders. What data is acceptable to collect and to what ends should it be applied? How much uncertainty, and potential bias, are we willing to accept in our predictive algorithms? How much autonomy do we want to concede to machines? Do we concern ourselves with the loss of jobs in our organization due to the automation of workflows? How much responsibility do we assume for users of our platforms? Do we view certain users as partners, employees, or independent contractors? What are our responsibilities to each? How do we wish to play in the "nonmarket" arena? Do we feel that all actions are fair game? Or are there limits to what we are willing to do to gain an edge or to win a stakeholder battle?

With regard to this last point, leaders need to recognize that your non-market strategy is part and parcel of your overall strategy. You cannot separate the two. With a group of colleagues, I contributed to a paper titled "CSR Needs CPR"—in other words, corporate social responsibility needs corporate political responsibility.[24] Beyond the clever title, the point was that companies that espouse a set of values on their company website and then engage in political behavior counter to those espoused values risk subjecting themselves to charges of hypocrisy. Even worse, as individuals and organizations, they demonstrate a lack of integrity—either a failure to live their espoused values or a failure to espouse their true values. It is beyond my role to tell you what your values should be, but recognize that your values matter. True leaders are authentic in their espoused values and work to build organizations and cultures that are aligned with them.

FRAMEWORK #6: PREPARING FOR DIGITAL IN SOCIETY

Once again, your nonmarket strategy should be aligned with your overall strategy. To begin, list the specific issues or challenges potentially created by your digital efforts in column one (see Figure 6.4). These could be issues around data privacy, antitrust concerns, autonomy, treatment of contingent workers, and so on. It is better to be more expansive than less. Identifying potential concerns early will better prepare you for dealing with them if they should arise.

For each issue identified, reflect on which organization stakeholders may feel an impact. List them in column two. Pay obvious attention to your primary stakeholders: customers, employees, supply chain partners. But do not forget secondary stakeholders such as activist groups, the government, and the media. Provide details on the potential impact for each stakeholder group and what they care about. The key is to be empathetic. Place yourself in the shoes of your stakeholders and reflect on their motivations and desires.

In column three, describe your strategic approach to the issue. Consider fight, flight, accommodate, partnering, and dual-purpose strategies. How do you plan to address the specific issue? Be cognizant of potential spillovers between actions. If you choose to fight on a particular issue, what impact will this have on other stakeholders and issues that you may face? If you choose to partner or accommodate on an issue, could this create positive spillovers across issues? Do not underestimate the "power of and." The best nonmarket strategy is often to look to jointly create value with your numerous stakeholders.

Last, and most important, how does your strategic approach align with your organizational values? Use column four to be explicit about alignment. Are these proposed actions consistent with your values and do they reinforce your overall competitive position? If there is misalignment between your values, your position, and your nonmarket strategy, you should reconsider your strategic approach. This may entail changes to your nonmarket strategy, or it may cause reflection on your broader competitive positioning. Think back to the Strategist's

Digital Issue	Stakeholder Impact	Strategic Approach	Values Alignment
Describe a specific issue or challenge potentially created by your digital efforts, e.g., data privacy, cybersecurity, antitrust, bias in algorithms, etc.	Who is affected by this issue and how? What do they care about?	How do you plan to address the issue? Consider fight, flight, accommodate, and dual-purpose strategies. Could include stakeholder engagement, lobbying, and other "nonmarket" strategies.	How does your strategy approach align with your organizational values and your competitive position?
See above.	See above.	See above.	See above.
See above.	See above.	See above.	See above.
See above.	See above.	See above.	See above.

Figure 6.4. Preparing for Digital in Society Framework.

Challenge: How do you align your values with the opportunities the market provides and the organization's aspired capabilities so as to establish competitive positions that create overall value? Your nonmarket strategy needs to be similarly aligned with all three dimensions of the Strategist's Challenge.

7

Transformation in the Digital Age

In 2006, two friends and former "Googlers," David Friedberg and Siraj Khaliq, founded WeatherBill. The concept behind their new venture was to leverage GPS, weather data, and analytics to create data resources for farmers so they could operate their farms more sustainably. The agriculture industry was at the epicenter of several environmental challenges. Topsoil depletion was having an impact on farm productivity and increasing the use of water and nitrogen-based fertilizer. Climate change loomed large, having an impact on the frequency of extreme weather and the availability of water. Farming both suffered from *and* contributed to the climate crisis, as it produced nitrous oxide—a potent greenhouse gas—through the use of fertilizers and by disturbing the soil. All the while, demand for food was increasing to meet a growing world population.

WeatherBill was one of an emerging set of AgTech businesses leveraging digital technology to help advance sustainable farming practices. Using a software-as-a-service business model, their decision support tool helped farmers monitor crop health and land use to optimize operations and maximize crop yield. Their technology complemented other digitally enabled technologies such as precision seed and liquid delivery systems from Precision Planting and automated tractors from John Deere. Their aspirations were nothing short of changing the world in a meaningful way.[1] Others agreed. WeatherBill acquired funding from Silicon Valley–based venture capitalists Khosla Ventures and Google Ventures, changed its name to the Climate Corporation in 2011, and eventually sold the business to Monsanto for $1.1 billion in 2013.[2] Today the company lives on, offering a suite of different digital solutions for precision agriculture and sustainable farming.

THE PROMISE OF DIGITAL TECHNOLOGY

Despite all the perils outlined in the previous chapter, the digital age holds great promise to address many of our most pressing global challenges. From climate change to poverty to discrimination, digitization and data analytics can provide new avenues for building solutions to our most seemingly intractable problems. Smart city initiatives in the United States and beyond look to leverage data to improve traffic flows, reduce crime, and identify those in need while providing a means to empower communities, increase transparency, and foster civic participation. Companies are leveraging digital technologies to address diverse challenges in health care, education, and finance—such as using AI to identify skin cancers, designing custom learning journeys for young at-risk students, and providing online financial services for the underbanked.

Climate change is one of our most pressing global challenges. Climate scientists warn that avoiding the worst impacts of global warming will require the reduction of global emissions of greenhouse gases to net zero by 2050, if not sooner—what is referred to as decarbonization.[3] Critical to a decarbonized future is a smart electrical grid that allows for the dynamic trading of electricity from distributed sources of power generation such as solar panels on homes and businesses and massive amounts of batteries and other forms of storage to help with the intermittency of renewables (the sun doesn't always shine, and the wind does not always blow). Digitization, IoT, and data analytics will all be critical to making such a smart grid system work. Companies such as C3 AI are working on providing just such a system.

Autonomous vehicles hold promise to radically change the way we consume transportation services and even how we organize our lives. There are the obvious benefits such as limiting accidents and reducing traffic congestion. Then there are second-order benefits such as reducing the number of cars on the road and helping catalyze the transformation to electric vehicles to eliminate emissions. More broadly, automation may change how we think about work-life balance, where to work, and where to live.

In the health care space, digital transformation is being hailed as nothing short of a revolution to health outcomes. Digital trackers, such as smart watches, are generating massive amounts of real-time data on your health. They have been used to identify heart conditions such as atrial fibrillation and to identify the potential onset of epileptic seizures. The COVID pandemic accelerated the adoption of telemedicine in the United States, highlighting the opportunity to expand health care to rural and disadvantaged communities. Social determinants of health data combined with patient histories and fueled by artificial intelligence are being adopted to provide more proactive care.[4]

In the education space, digitization is opening access to educational opportunities to marginalized peoples and communities. In the early days of my massively open online course on the Coursera platform, I had a cohort of Mongolian students work together with a Peace Core volunteer to form a study group while taking the course. A group of women in a US shelter for those who had suffered abuse were taking my course to help them launch entrepreneurial ventures to get themselves back on their feet. In the Middle East, a group of Israeli and Palestinian students were taking my course, in part to help build detente between the two groups.

I could go on and on. Digitization has the potential to transform virtually every industry, create new opportunities for value creation, and, on many occasions, provide opportunities to better address pressing societal issues. The end of privacy, cybersecurity, bias in algorithms, social and political discord spurred by technology, AI running rampant, and the rise of technology behemoths are all serious issues that require significant and dedicated attention. However, these risks should not obscure the opportunities that digitization and digital transformation create. The question before us all is, How do we best leverage digital technology to create value and improve the world?

THE JOURNEY TO TRANSFORMATION

The question above highlights the importance of being intentional and thoughtful in your digital transformation efforts. History is littered with well-intentioned but ultimately misguided efforts at disruption and transformation. The digital age is certainly no different. To begin your digital journey, start with the Strategist's Challenge and an articulation of your

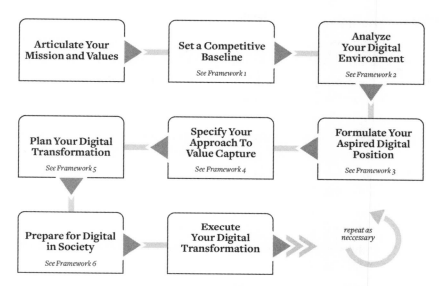

Figure 7.1. A Roadmap for Digital Transformation.

mission and values (see Figure 7.1). What is your north star that guides your actions and decisions? What do you and your various stakeholders expect of your organization? How do you define and measure success, individually and as an organization? What values are immutable and define the way you will play the game, the rules that you will play by?

Next, analyze the opportunities presented by the digital age. Start with the basics (see Framework 1). What are some of the trends having an impact on your sector? Are there demographic, sociocultural, political-legal, technological, macroeconomic, and global trends you should be tracking? Identify your competition; not only the organizations with which you currently compete for attention and resources, but potential new competitors who may enter your domain and compete with you in the future. What are the strengths and weaknesses of these current and potential competitors? Reflect on your own capabilities: What are the underlying people, processes, and systems that allow you to create value? Are these capabilities well aligned to deliver on a consistent value proposition? Are these capabilities durable and hard to imitate, and will they ultimately provide a sustainable competitive advantage?

Analyze the nature of competition in your sector (see Framework 2). In what stage are you in the industry life cycle? Have new technologies caused

132 STRATEGY IN THE DIGITAL AGE

a significant disruption that is leading to a new emergent epic? Perhaps digital disruption has begun apace, with new technologies and business models gaining traction and growing in the marketplace. Perhaps the industry is more mature, even facing decline and potential new disruption. Similarly, where are you in your organization's life cycle? Are you a nascent venture, a growth company, a mature competitor, or a troubled player? Consider the implications of these life cycles on the nature of competition and your ability to compete.

Reflect on how the digital age may be changing the basis of competition in your sector. Consider the industry archetype today and in the future. Does digitization open the possibility of a dominant platform emerging that leverages network externalities and creates a winner-take-all market? Perhaps digitization is creating a more segmentable market with multiple platform plays available? Or perhaps there will be defensible positions created that leverage intellectual property or other barriers to competition? Last, perhaps a digital transformed sector will be a more competitive marketplace in which competitive advantage does not last long, and success relies on complementary capabilities?

With an understanding of the emerging marketplace in the digital age, you can begin to articulate an aspired competitive position (see Framework 3). What is your desired general competitive position? For example, do you aim to be a low-cost leader or a high-end differentiated player? Where do you play in the value chain, recognizing that digitization often deconstructs the value chain, making available numerous positions and business models? Who is your target customer and what is your value proposition for these customers? How can you leverage data to enhance this value proposition and to create long-term continuous relationships? Next, articulate your business model. Do you plan to use a traditional fee-for-service model? A subscription model like a software-as-a-service offering? Perhaps an advertising, freemium model, or some combination of the above?

Be explicit about how you will be able to defend this position from rivals who are perhaps looking to establish a similar position. Do you plan to leverage network externalities to build a defense against imitators? Perhaps you will build in switching costs to help lock in customers to your offering. Or leverage learning curves and interdependencies to establish a dominant position. Maybe you will capture branding advantages or secure cornered resources such as patents and copyrights to secure your position. The key is to be realistic. You may not be best positioned to achieve every desired position. How do your current capabilities provide a basis to build a competitive advantage in the digital age?

With an aspired competitive position articulated, you need to consider the evolving set of capabilities you will need to deliver and defend this position (see Framework 4). Articulate your needs across your value chain, including upstream and downstream. Formulate a strategy for delivering those needs, for example, whether you intend to build them internally, acquire them externally, and/or partner with others. Be explicit on how you will appropriate value from these build, buy, and buddy strategies. Which of these capabilities could be the basis for a competitive advantage?

With these capabilities defined, you are now ready to consider the specific applications to develop as part of your digital transformation (see Framework 5). Consider the Digital Transformation Stack. Start with your digital strategy. Which digital applications help deliver the capabilities needed to achieve a desired competitive position? These may relate to your core product offerings and/or underlying operational needs in finance, HR, and marketing, for example. Take a hypothesis-driven approach to application development. Identify core assumptions. What are the data needs underlying each application? What core digital infrastructure needs do you have to have in order to be able to process data that will feed these desired applications? Which data analytic skills are necessary to leverage this data for productive uses? What are the value assumptions behind the application? How will this application deliver value and to whom? What must be true for you to execute this application? What people, processes, and systems are required of the organization? Will this application provide a competitive advantage and if so, how? Will it help solidify your aspired competitive position in the market?

Last, but certainly not least, consider the impact your efforts may have on broader society (see Worksheet 6). Articulate potential issues or challenges created by your digital efforts. For each one, consider what impact the issue may have on various stakeholders of the firm. What do these stakeholders care about and value? For each issue, devise a strategic approach. How do you plan to address the issue? Consider fight, flight, accommodate, partnership, and dual-purpose strategies such as stakeholder engagement, lobbying, and other "nonmarket" approaches. Reflect on how your approach aligns with your organization's values and your aspired competitive position. Are these misaligned? If so, what does this say about your strategy or your approach? Do you need to adjust to align your values, position, and actions?

This last question is critical, as it highlights the reciprocal nature of your digital transformation journey. Strategy is a process of constant reflection as markets and technology advance, competitors rise and fall as they jockey for positions, and societal norms and expectations evolve. Strategy in the digital age is no different. Successful digital transformations require constant analysis and adjustments in fast-moving environments that are continually evolving.

THE FUTURE OF DIGITAL

Ultimately, your digital transformation is a journey with no end. As discussed in the first chapter, the exponential growth in the core digital technologies of processing power, bandwidth, and storage could lead to a world radically different from what we see today in a relatively short period of time. Imagine more data generated next year than all previous years combined, 30x growth in computing power over the next decade, and artificial intelligences achieving some version of general intelligence in our lifetimes. The first of these two are highly likely possibilities, while the third is hotly debated.

New digital technologies and concepts are constantly emerging. Facebook changed its corporate name to Meta in 2021 to reflect its aspirations to be a leader in the emerging metaverse. The metaverse is a vision for a new way of engaging with the Internet utilizing 3D rendering and virtual reality. Rather than a flat, two-dimensional interface, the Internet would become an immersive experience in which we utilize 3D goggles and haptic suits to engage with the digital world. While arguably such a future is already upon us in video gaming, proponents of the metaverse see it as much more expansive and potentially disruptive. They see the metaverse as the way we will connect with one another (replacing existing social media platforms), shop (evolving the e-commerce experience), and even learn (disrupting universities and advancing online education).

Some characterize this as the third evolution of the web, what is sometimes referred to as Web3. The first evolutionary period was that of the browser and is represented by the likes of Google and Amazon. The second period featured social media and interactivity and is represented by the likes of Facebook/Meta and Twitter. Advocates of a third epic see the metaverse leveraging other emerging technologies such as blockchain. The metaverse will be a place where people can buy and trade digital goods and services using cryptocurrencies, free of the constrictors of the physical world and independent of current institutions and third-party intermediaries (like regulated banks). Digital assets will be assigned to owners as nonfungible tokens, or NFTs, and verified in a distributed blockchain. Likely you will own a truly unique avatar, unable to be replicable by others, to navigate this new digital world.

The market for NFTs has exploded in recent years. In March 2021, the digital graphic artist Beeple auctioned off a collage of his digital work through Christie's Auction House for $69 million.[5] For the buyers, their $69 million got them a nonfungible token recorded on the Ethereum blockchain so as to assert ownership (and bragging) rights. Interestingly, they did not get formal copyright. The Bored Apes Yacht Club is a collection of ten thousand digital images of cartoon apes, each featuring a unique combination of outfits, expressions, and hairstyles. Bored Apes NFT owners get exclusive rights to their digital ape, perhaps to be used as their profile picture on Twitter, and access to Bored Ape Yacht Club social events online. In April 2022, the average price of one Bored Ape NFT was $500,000, with the rarest of Apes—that featured unlikely combinations of attributes—fetching over $3 million.[6] In a sign of its popularity, the Bored Apes Yacht Club attracted celebrities such as late-show host Jimmy Fallon and NBA superstar Steph Curry.[7]

The metaverse may very well blur the lines between the real and digital worlds. Augmented reality (AR) provides a potential interface between the metaverse and our engagement with the physical environment. In July 2016, Nintendo in partnership with AR venture Niantic released the Pokemon Go augmented reality game in which players discover digital Pokemon in physical spaces and objects using their mobile phones. It was a viral sensation and highlighted the potential of AR. Companies such as Amazon are getting in on the action, creating AR experiences that allow you to "try on" shirts and shoes before you purchase them online.

Coders are actively building digital twins—exact digital replicas of real-world places and architectures. Bentley Systems partnered with

Singapore to create a digital twin of the entire city-state. Similar projects are in the works for major cities such as New York. Companies are creating digital twins of their factories to be able to simulate outcomes, identify efficiencies, and troubleshoot potential problems. Leveraging the Internet of Things, they are connecting machines and sensors to generate continuous real-time data on their processes that are fed into their digital twins, creating a dynamic exchange between the physical and digital worlds.

Of course, the metaverse is not really a new concept. As is often the case with technology, science fiction has been advancing various visions of a metaverse for decades. Neal Stephenson is largely credited with coining the term "metaverse" in his 1992 book *Snow Crash*. The 1999 film *The Matrix* featured a metaverse that serves to entrap a humanity blissfully unaware that they are "living" in a simulation and enslaved in the real world. The 2011 book *Ready Player One*, by Ernest Cline, and subsequent movie provide a detailed vision for what a commercial metaverse could look like, and some of the disconcerting social implications of an addictive, all-consuming immersive 3D digital world.

So is the metaverse, for good and for bad, our destiny? Maybe, but maybe not. In 2003, Linden Labs launched the virtual world Second Life. Second Life was an early attempt to establish a metaverse that became a viral sensation. In 2006, schoolteacher Ailin Graef (avatar name Anshe Chung) made over $1 million developing virtual real estate projects in Second Life,[8] her avatar even landing on the cover of *BusinessWeek*. Today, Second Life still exists but has largely faded from the public eye. A concept before its time? Perhaps. Or maybe the hype around the metaverse exceeds an individual's desire to use an immersive 3D environment to chat with friends and buy toiletries online. Maybe the two-dimensional Internet is good enough for us most of the time.

As you read this book, it is likely that some of the technologies I reference will have already failed and new ones, wholly unforeseen, will have emerged. In May 2022, spurred by the collapse of the algorithmic stable coin TerraUSD, the general market for cryptocurrencies fell 40 percent, destroying $300 billion in asset value in a few days.[9] Does this spell the end of cryptocurrency? Perhaps. I certainly have my own doubts of the need for and viability of cryptocurrencies. Bitcoin is still overwhelmingly used as a speculative investment asset and not a method of conducting financial transactions. History will be the judge.

Consider artificial intelligence. AI advancements over the past decade have been truly astonishing. In May 2022, the OpenAI project released the DALL-E 2 visual recognition and creation bot. With a couple of simple text-based prompts, DALL-E 2 will draw a corresponding digital image. Ask to see an otter in a top hat on a bicycle and the AI is shockingly capable of delivering. You can even specify your artwork in specific styles, say Japanese anime or French impressionist styles, and DALL-E 2 will comply. AIs are also creating music. An AI called "Aiva" (Artificial Intelligence Virtual Artist) can create classical music. An album of Aiva's works has been released and songs from Aiva have appeared on movie soundtracks.[10]

Artificial-intelligence-driven autonomy has advanced much quicker than many had predicted. Not too long ago, analysts were citing truck drivers as one of the few jobs to be relatively protected from automation. In 2022, there is hope to have fully autonomous trucks plying our highways in the next five years. We are already seeing the deployment of fully autonomous vehicles, from pizza delivery bots on college campuses to John Deere tractors in the fields. So is it only a short matter of time before we are all zipping around town in a fully autonomous car? Elon Musk seems to think so, continually predicting that Level 5 autonomy for autos is coming "next year." Others are less optimistic, and some are downright pessimistic.

What about even more advanced realizations of AI? How long until we have humanoid robots like those portrayed in science fiction, from Rosie on *The Jetsons* to Data on *Star Trek: The Next Generation*? (Or, heaven help us, how long until the Terminator roams the streets?) Again, I don't know. And neither does anyone else with certainty. That is the nature of technology. Forecasting technology trends is notoriously difficult. In 1943, Thomas Watson, president of IBM, predicted that "I think there is a world market for maybe five computers."[11] He envisioned a world in which a handful of supercomputers would provide all our computational needs, totally missing the microcomputer revolution and the billions of computing devices that we now carry around with us.

THE POWER OF PURPOSE

What is clear is that your digital strategy must constantly adjust and evolve to fit an ever-changing world. Digital transformation can be difficult to manage for intrepid entrepreneurs and established incumbents alike—fundamentally disruptive to markets and competition, creating both opportunities and risks. To navigate the ever-changing currents of digitization requires an ability to be both deliberate and nimble, to be strategic in the true sense of the word. Successful organizations need to identify and secure valuable competitive positions at the intersection of their values, the opportunities provided by the market, and the organization's unique and hard-to-imitate capabilities. This is a dynamic process, responding to evolving market conditions, requiring constant innovation, adjusting capabilities and offerings to create and capture value.

To this end, managers must invest in their digital infrastructure and analytic capabilities. They need to organize and lead efforts to build out digital applications driven by their digital strategy. More broadly, successful leaders prepare their organizations to be able to nimbly adjust to the ever-changing technological and competitive landscape. They demonstrate humility, knowing that the future is uncertainty and technology continues to improve exponentially. They are thoughtful not only about the narrow marketplace, but about the broader institutional environment in which business operates, devising nonmarket strategies for addressing the inevitable concerns and perils of digitization.

And, always, successful leaders are powered by purpose. They are guided by their values that define where and how they are willing to play. They are inspired by the impact they wish to have on the world. And they use this purpose and vision to inspire others to be partners in this journey. The digital age may be scary, but it can also be transformational. Digital technologies hold promise to help address the world's most pressing challenges even as they create new concerns and issues. The question is, How will you and your organization leverage digital transformation to transform the world for the better?

Glossary

AB testing: The use of true experiments featuring a treatment group and a control group.

Absorptive capacity: A firm's ability to recognize the value of new information, assimilate it, and apply it to commercial ends, as defined by Wes Cohen and Dan Levinthal.

Additive manufacturing: Computer-controlled manufacturing process that applies layers of material to create a 3D product.

Agile development: An iterative approach to software development that builds applications incrementally.

Algorithm: Procedure for solving a mathematical problem in a finite number of steps that frequently involves a repetition of an operation. For computers, it's the set of rules that machines follow to reach an end goal. (Source: *Merriam Webster's Dictionary*, 2021)

Antitrust: Relating to legislation preventing or controlling trusts or other monopolies, with the intention of promoting competition in business. (Source: *Oxford Dictionary*)

Application Programming Interfaces: Instructions for interacting with a specific software.

Appropriation: The capturing of value from an innovation, either as an innovator, competitor, customer, or value-chain partner.

Artificial intelligence (general): The ability of a machine to not only carry out specific tasks but also to possess other human brain characteristics, such as self-awareness, reasoning, problem solving, and planning for the future. Also known as "strong" AI, these systems would be indistinguishable from human intelligence.

Artificial intelligence (narrow): The ability of a machine to perform narrowly defined tasks usually performed by human beings. Narrow AI encompasses many digital technologies including, but not limited to, machine learning, voice recognition, and natural language processing.

Augmented reality: An interactive experience in which engagement in the real world is enhanced by digital artifacts.

Autonomy: The capacity to self-govern and make decisions independently. For AI, this means that the system can make decisions without direct human interference.

Blockchain: A digital ledger of peer-to-peer transactions that is duplicated across a series of computer systems across a decentralized network.

Business architecture: The set of players that interact together to deliver a product or service.

Business Model Canvas: Invented by Swiss business theorist and entrepreneur Alex Osterwalder, it is a strategic management template for creating business models using the following inputs: core value proposition, targeted customers, key activities, resources and partners, cost structure, and revenue streams.

Cloud computing: The hosting and delivery of on-demand data storage and other services through the Internet using a network of remote computer servers.

Coach behavior: Experiences in which predicted algorithms and behavior nudges are used to change customer behavior.

Collaborative filtering: Technique used by AI recommender systems to make predictions about a user's interests that are based on the behavior of multiple, similar users.

Competitive advantage: Factors that allow a company to produce goods or services better or more cheaply than its rivals. These factors allow the productive entity to generate more sales or superior margins compared to its market rivals. (Source: Investopedia.com)

Competitive Life Cycle: The evolution of competition in a market as a technology advances and matures.

Complementary capability: An activity and/or ability that enhances the appropriation of value from an innovation, for example, the ability to manufacture an innovative product.

Connected strategies: Strategies that turn occasional, sporadic transactions with customers into long-term continuous relationships, as defined by Nicolaj Siggelkow and Christian Terwiesch.

Cross-side network effects: The value of a service increases for one user group with the addition of new users to a different user group within the same network. This is often observed in online platform businesses, when two or more user groups are exchanging products or services with one another.

Cryptocurrency: A digital currency designed to work as a medium of exchange through a computer network that is not reliant on any central authority, typically using a blockchain as ledger of transactions.

Cybersecurity: The act of protecting computers, networks, and other digital assets from cyberattacks.

Data dashboard: A management tool that allows for the real-time tracking, analysis, and display of data essential for monitoring business performance.

Data lake: Central repository for storing structured and unstructured data.

Data mining: Sorting through and analyzing raw datasets to identify trends and patterns.

Data science: Field of study that combines multiple fields, including statistics, scientific methods, artificial intelligence, and data analysis, to extract value from data. Those who practice data science are called data scientists, and they combine a range of skills to analyze data collected from the web, smartphones, customers, sensors, and other sources to derive actionable insights. (Source: Oracle.com)

Decision forest (random): An algorithm that randomly creates and merges multiple decision trees to produce a more accurate prediction.

Demand economies of scale: The increasing value provided by a product or service within a network as the number of users increases.

Descriptive analytics: The examination of data to summarize and describe what has happened in the past.

Design thinking: A problem-solving approach that is human-centered, possibility-driven, option-focused, and iterative.

Diagnostic analytics: The examination of data to determine why something happened.

Digital disruption: Within markets, the radical shift in value provided by goods and services, driven by digital technologies.

Digital transformation: The process of leveraging digital technologies to innovate products, services, and business processes to meet changing business and market requirements.

Digital twins: Exact digital replicas of real-world places and architectures.

Digitization: The conversion of analog to digital form in which the data itself is not changed.

Disruptive innovation: An innovation that transforms the market by displacing incumbent technologies and shaking up the established competitive landscape.

Economies of scale: The cost reductions realized as the volume of goods and services increases. Typically expressed as cost per unit produced.

Enabling technologies: Technologies that enable the rapid development of derivative technologies and solutions.

Fintech: Term used to describe new technologies that seek to improve and automate the delivery and use of financial services. (Source: Investopedia.com)

Freemium model: Type of business model that relies on a two-tiered system in which users can access services for free but then pay a premium for additional content or services.

General purpose technologies: Innovations that have the potential to have an impact on the broader economic system and society at large. Examples include the Internet and electricity.

Horizontal diversification: Adding new products to an existing product line, or introducing brand new product lines, to serve existing customer needs. Often these new products provide an opportunity for companies to shift into new, adjacent markets.

Innovation ecosystem: A space that allows for the collective effort between market actors and other stakeholders to bring an innovation to scale. Actors may include policymakers, universities, private corporations, entrepreneurs, and investors, all of which bring a specific skill set and added value to the system.

Intangible assets: Assets that are not physical in nature. Brands and company culture are examples of intangible assets. (Source: Investopedia.com)

Interdependency: The dependence of two or more activities on each other when the value of one activity or asset is dependent on another.

Internet of Things (IoT): The network of physical objects—"things"—that are embedded with sensors, software, and other technologies for the purpose of connecting and exchanging data with other devices and systems over the Internet. (Source: Oracle.com)

Machine learning: A branch of AI and computer science that focuses on the use of data and algorithms to imitate the way that humans learn, gradually improving accuracy. (Source: IBM.com)

Metaverse: An extensive online world characterized by 3D rendering and virtual reality.

Metcalf's Law: As defined by Robert Metcalf, founder of 3Com, the number of connections in a network grows nonlinearly as the number of nodes, or users, increases.

Modularity: Systems composed of units that are designed independently but still function as an integrated whole, as defined by Carliss Baldwin and Kim Clark.

Moore's Law: Named in recognition of Intel founder Gordon Moore, the forecast that the number of transistors on a microchip doubles roughly every two years.

Natural language processing: The branch of AI and computer science concerned with giving computers the ability to understand text and spoken words in much the same way human beings can. (Source: IBM.com)

Net neutrality: The idea, principle, or requirement that Internet service providers should or must treat all Internet data as the same regardless of its kind, source, or destination. (Source: Merriam-Webster.com)

Network externality: A change in the benefit, or surplus, that an agent derives from a good when the number of other agents consuming the same kind of good changes, as defined by S. J. Liebowitz and Stephen Margolis.

Niche player: In terms of market competition, a niche player chooses to compete in a narrower segment of the market, offering specialized goods favored by these targeted customers.

Non-fungible token (NFT): A way of exerting ownership rights of digital assets, typically using a blockchain to record those rights.

Open architecture: Software or technology that is nonproprietary and open to other users to add or change components.

Open innovation: Management practice that allows for collaboration with actors outside of an organization to accelerate innovative activities within that organization.

Partnering model: Business model in which a facilitator to a transaction, such as a platform, charges a percentage fee for any funds exchanged; typical of a two-sided market, when the customer pays a fee to the other party of exchange.

Patent fence strategy: The building up of a large portfolio of patents to serve as a fence-line against those who may sue for patent infringement on their own patents.

Platform: A hardware and/or software architecture that hosts other applications or programs.

Predictive analytics: Advanced analytics that make predictions about future outcomes using historical data combined with statistical modeling, data-mining techniques and machine learning. (Source: IBM.com)

Prescriptive analytics: Advanced analytics that use optimization and simulation algorithms to advise on possible outcomes and recommend a course of action.

Primary stakeholders: Market actors with which a company has a direct transactional relationship. Examples include suppliers, employees, and customers.

Quality differentiator: In terms of market competition, the quality differentiator incurs higher costs to offer differentiated goods and services for which customers are willing to pay a price premium.

Red Queen Effect: The idea that businesses must continuously evolve and adapt to merely keep up in a competitive market.

Regression analysis: A statistical method for determining the relationship between a dependent and independent variable or variables.

S-curve: Presented as a visual to describe the technology cycle, the S-curve illustrates the rate in which a new technology diffuses into the market, from emerging to mature, dominant design.

Same-side network effects: The value of a service increases for one user group with the addition of new users to that side of a multisided network.

Secondary stakeholders: Market actors who have an impact on and are affected by the actions of one company, either directly or indirectly. Examples include activists, regulators, the media.

Software as a service (SaaS): Centrally located, subscription-based model for distributing software to users.

Solow's Paradox: Based on work by economist Robert Solow, the phenomenon in which investment into IT systems continues to increase yet productivity does not.

Switching costs: The costs incurred by a customer for switching to an alternative supplier.

Tangible assets: Assets in physical form. Examples include cash, buildings, vehicles, and product inventory.

Technological singularity: A future in which human intelligence is overtaken by artificial intelligence.

Technology stack: Set of digital technologies used to build web or mobile applications, in which each layer is built on top of another layer, creating a "stack."

Transaction costs: Costs incurred to secure an economic transaction beyond the direct costs of purchasing a good or service.

Two-sided market: An intermediary platform that facilitates the exchange of goods and services between sellers and buyers.

Value chain: The full set of activities performed by a company to create a product or service (for example, sourcing, manufacturing, assembly).

Vertical integration: Owning elements of the supply chain, such as manufacturing and distribution. An example is an electric utility that owns power generation, supply, and distribution to customers.

Virtual reality: A computer-generated environment in which users feel immersed in their surroundings.

Web3: A loosely defined vision for third-generation interaction with the Internet featuring virtual reality, 3D rendering, and extensive use of blockchain technology.

Winner-take-all market: A market in which a single company dominates the gains from trade within that market.

Wireframe: A simple, two-dimensional illustration of a web page, app interface, or product layout.

Notes

Chapter 1

1. James Estrin, "Kodak's First Digital Moment," blog, *The New York Times*, August 12, 2015, https://lens.blogs.nytimes.com/2015/08/12/kodaks-first-digital-moment/.

2. Xavier Markl, "Hamilton PSR, Reviving the First-Ever Digital Watch," Hands-on, Monochrome.com, March 18, 2020, https://monochrome-watches.com/hamilton-psr-reedition-hamilton-pulsar-first-digital-watch-review-price/.

3. Thomas Siebel, "Why Digital Transportation Is Now on the CEO's Shoulders," *McKinsey Quarterly*, December 14, 2017, https://c3.ai/tom-siebel-ceos-transform-disappear/.

4. S. Patrick Viguerie, Ned Calder, and Brian Hindo, "2021 Corporate Longevity Forecast," Innosight https://www.innosight.com/insight/creative-destruction/May 2021.

5. Robert E. Siegel and Julie Makinen, "C3 IoT: Enabling Digital Industrial Transformation," Case No. SM307, Stanford Graduate School of Business, 2018, https://www.gsb.stanford.edu/faculty-research/case-studies/c3-iot-enabling-digital-industrial-transformation.

6. Singularity.com, "Magnetic Data Storage (Bits per Dollar) 1952—2004," accessed September 20, 2021, http://www.singularity.com/charts/page76.html.

7. Jakob Neilsen, "Neilsen's Law of Internet Bandwidth," Neilson Normal Group website, April 4, 1998 (updated September 27, 2019), https://www.nngroup.com/articles/law-of-bandwidth/.

8. Christo Petrov, "25+ Impressive Big Data Statistics for 2021," Techjury blog, accessed June 2, 2021, https://techjury.net/blog/big-data-statistics/#gref.

9. Ibid.

10. Radoslav Ch., "37 Heavenly Cloud Computing Statistics for 2021," Techjury blog, accessed June 3, 2021, https://techjury.net/blog/cloud-computing-statistics/#gref.

11. Ibid.

12. Nick G., "How Many IoT Devices Are There in 2021? [All You Need To Know]," Techjury blog, accessed June 2, 2021, https://techjury.net/blog/how-many-iot-devices-are-there/#gref.

13. Ibid.

14. Ibid.

15. Greg Kumparak, "Google Is Lowering Next Camera Quality to Conserve Internet Resources," TechCrunch, April 14, 2020, https://techcrunch.com/2020/04/14/google-is-lowering-nest-camera-quality-to-conserve-internet-resources/.

16. Fred Lambert, "Tesla Is Collecting Insane Amount of Data from Its Full Self-Driving Test Fleet," Electrek, October 24, 2020, https://electrek.co/2020/10/24/tesla-collecting-insane-amount-data-full-self-driving-test-fleet/.

17. Bob O'Donnell, "How Fast Will 5G Really Be?" *Forbes*, November 19, 2019, https://www.forbes.com/sites/bobodonnell/2019/11/19/how-fast-will-5g-really-be/?sh=1004b5cd5cf3.

18. M. Attaran, "The Impact of 5G on the Evolution of Intelligent Automation and Industry Digitization," *Journal of Ambient Intelligence and Humanized Computing* (February 21, 2021), https://doi.org/10.1007/s12652-020-02521-x; https://link.springer.com/article/10.1007/s12652-020-02521-x.

19. Eric Rosenberg, "How Much Energy It Takes to Power Bitcoin," The Balance, accessed June 3, 2021, https://www.thebalance.com/how-much-power-does-the-bitcoin-network-use-391280.

20. Ibid.

21. Simplilearn, "What Is Blockchain Technology and How Does It Work?" accessed June 3, 2011, https://www.simplilearn.com/tutorials/blockchain-tutorial/blockchain-technology.

22. Ron Adner, Phanish Puranam, and Feng Zhu, "What Is Different About Digital Strategy? From Quantitative to Qualitative Change," *Strategy Science* 4, no. 4 (December 10, 2019), https://pubsonline.informs.org/doi/full/10.1287/stsc.2019.0099.

23. Neelam Teyagi, "How Spotify Uses Machine Learning Models," Analytic Steps, accessed June 3, 2021, https://www.analyticssteps.com/blogs/how-spotify-uses-machine-learning-models.

24. Klaus Schwab, "The Fourth Industrial Revolution: What It Means, How to Respond," World Economic Forum, January 14, 2016, https://www.weforum.org/agenda/2016/01/the-fourth-industrial-revolution-what-it-means-and-how-to-respond/.

25. McKinsey & Company, "How COVID-19 Has Pushed Companies Over the Technology Tipping Point—and Transformed Business Forever," October 5, 2020, https://www.mckinsey.com/business-functions/strategy-and-corporate-finance/our-insights/how-covid-19-has-pushed-companies-over-the-technology-tipping-point-and-transformed-business-forever#.

26. Ajay Agrawal, Joshua Gans, and Avi Goldfarb, *Prediction Machines* (Boston: Harvard Business Review Press, 2018).

27. Edward D. Hess and Katherine Ludwig, *Humility Is the New Smart: Rethinking Human Excellence in the Smart Machine Age* (Oakland, CA: Berrett-Koehler, 2017).

28. Peter F. Drucker, "The Theory of Business," *Harvard Business Review Magazine* (September-October 1994), https://hbr.org/1994/09/the-theory-of-the-business.

29. Jared Harris and Michael Lenox, *The Strategist's Toolkit* (Charlottesville, VA: Darden Business, 2013).

Chapter 2

1. Todd W. Schneider, "Taxi and Ride Hailing Usage in New York City," toddwschneider.com, accessed September 15, 2021, https://toddwschneider.com/dashboards/nyc-taxi-ridehailing-uber-lyft-data/.

2. Mansoor Igbal, "Uber Revenue and Usages Statistics (2021)," Business of Apps, August 5, 2021, https://www.businessofapps.com/data/uber-statistics/.

3. Ibid.

4. Geoffrey G. Parker, Marshall W. Van Alstyne, and Sangeet Paul Choudary, *Platform Revolution: How Networked Markets Are Transforming the Economy and How to Make Them Work for You* (New York: W. W. Norton, 2016), 3.

5. S. J. Liebowitz and Stephen E. Margolis, "Network Externalities (Effects)," accessed July 8 , 2021, https://personal.utdallas.edu/~liebowit/palgrave/network.html.

6. Carl Shapiro and Hal Varian, *Information Rules: A Strategic Guide to the Network Economy* (Boston: Harvard Business Review Press, 1998).

7. Richard Lipsey and Kenneth Carlaw, *Economic Transformations: General Purpose Technologies and Long Term Economic Growth* (Oxford, UK: Oxford University Press, 2006).

8. Statcounter, "Search Engine Market Share Worldwide," August 2021, https://gs.statcounter.com/search-engine-market-share.

9. Ibid.

10. Statcounter, "Search Engine Market Share in China," August 2021, https://gs.statcounter.com/search-engine-market-share/all/china.

11. Lauren Thomas, "As e-Commerce Sales Proliferate, Amazon Holds On to

Top Online Retail Spot," CNBC, June 18, 2021, https://www.cnbc.com/2021/06/18/
as-e-commerce-sales-proliferate-amazon-holds-on-to-top-online-retail-spot.
html.

12. GMA, "Chinese e-Commerce Platforms in 2021: Which Is the Most Suitable for Your Brand?" GMA China e-Commerce, February 25, 2021, https://ecommercechinaagency.com/great-chinese-online-marketplaces-for-e-commerce/.

13. Statcounter, "Desktop Operating System Market Share Worldwide," August 2021, https://gs.statcounter.com/os-market-share/desktop/worldwide.

14. Ibid.

15. Statcounter, "Mobile Operating System Market Share Worldwide," August 2021, https://gs.statcounter.com/os-market-share/mobile/worldwide.

16. Synergy Research Group, "Global Quarterly Market Share of Cloud Infrastructure Services from 2017 to 2021, by Vendor (Q2 2021)," Statista, July 2021, https://www.statista.com/statistics/477277/cloud-infrastructure-services-market-share/.

17. Strategy Analytics, "Leading Ride-Hailing Operators Worldwide as of November 2019, Based on Market Share," Statista, August 2020, https://www.statista.com/statistics/1156066/leading-ride-hailing-operators-worldwide-by-market-share/.

18. Amazon, "Share of Paid Units Sold by Third-Party Sellers on Amazon Platform as of 2nd Quarter 2021," Statista, July 2021, https://www.statista.com/statistics/259782/third-party-seller-share-of-amazon-platform/.

19. Parker, Van Alstyne, and Choudary, *Platform Revolution*, 44.

20. Companiesmarketcap.com, "Market Capitalization of Microsoft," September 2021, https://companiesmarketcap.com/microsoft/marketcap/.

21. Peter Gratzke and David Schatsky, "Banding Together for Blockchain: Does It Make Sense for Your Company to Join a Consortium?" Deloitte Insights, August 16, 2017, https://www2.deloitte.com/us/en/insights/focus/signals-for-strategists/emergence-of-blockchain-consortia.html.

22. "The Global Startup Ecosystem Report 2020," accessed June 1, 2022, https://startupgenome.com/report/gser2020.

23. Richard A. D'Aveni, *Hypercompetition: Managing the Dynamics of Strategic Maneuvering* (New York: Free Press, 1994).

Chapter 3

1. Kyle Wiggers, "Waymo's Autonomous Cars Have Driven 20 Million Miles on Public Roads," Venture Beat, January 6, 2020, https://venturebeat.com/2020/01/06/waymos-autonomous-cars-have-driven-20-million-miles-on-public-roads/.

2. Tom Pritchard, "Apple Car: Everything We Know So Far," Tom's Guide,

July 20, 2021, https://www.tomsguide.com/news/apple-car-everything-we-know-so-far.

3. Josh Lowensohn, "Uber Gutted Carnegie Mellon's Top Robotics Lab to Build Self-Driving Cars," The Verge, May 19, 2015, https://www.theverge.com/transportation/2015/5/19/8622831/uber-self-driving-cars-carnegie-mellon-poached.

4. Carliss Y. Baldwin and Kim B. Clark, "Managing in an Age of Modularity," *Harvard Business Review* (September-October 1997), https://hbr.org/1997/09/managing-in-an-age-of-modularity.

5. David L. Rogers, *The Digital Transformation Playbook: Rethink Your Business for the Digital Age* (New York: Columbia Business School Publishing, 2016).

6. Nicolaj Siggelkow and Christian Terwiesch, *Connected Strategy: Building Continuous Customer Relationships for Competitive Advantage* (Boston: Harvard Business School Press, 2019), ix.

7. Megan Graham and Jennifer Elias, "How Google's $150 Billion Advertising Business Works," CNBC, May 18, 2021, https://www.cnbc.com/2021/05/18/how-does-google-make-money-advertising-business-breakdown-.html.

8. Rishi Iyengar, "Here's How Big Facebook's Ad Business Really Is," CNN, July 1, 2020, https://www.cnn.com/2020/06/30/tech/facebook-ad-business-boycott/index.html.

9. Michael Porter, *Competitive Advantage: Creating and Sustaining Superior Performance* (New York: Free Press, 1986).

10. Jeremy Bowman, "Does Stitch Fix Have a Competitive Advantage?" The Motley Fool, June 3, 2018, https://www.fool.com/investing/2018/06/03/does-stitch-fix-have-a-competitive-advantage.aspx.

11. Hamilton Helmer, "7 Powers: The Foundations of Business Strategy," Deep Strategy, LLC, October 27, 2016.

12. It is interesting to note that technically Stanford University owned the Page Rank Patent and licensed it to Google. The Page Rank Patent expired on September 24, 2019.

13. Dorothy Leonard, "Core Capability and Core Rigidities: A Paradox in Managing New Product Development," *Strategic Management Journal* 13 (1992): 111–125, 10.1002/smj.4250131009.

Chapter 4

1. Grandview Research, "Cyber Security Market Size, Share & Trends Analysis Report by Component, by Security Type, by Solution, by Services, by Deployment, by Organization, by Application, by Region, and Segment Forecasts, 2021–2028 (Report Overview)," accessed September 6, 2021, https://www.grandviewresearch.com/industry-analysis/cyber-security-market.

2. Gartner, "Gartner Forecasts Worldwide IT Spending to Reach $4 Trillion in 2021," Press Release, April 7, 2021, https://www.gartner.com/en/newsroom/press-releases/2021-04-07-gartner-forecasts-worldwide-it-spending-to-reach-4-trillion-in-2021.

3. Robert M. Solow, "We'd Better Watch Out," *New York Times* Book Review of Stephen Cohen and John Zysman, *Manufacturing Matters: The Myth of the Post-Industrial Economy*, July 12,1987, http://www.standupeconomist.com/pdf/misc/solow-computer-productivity.pdf.

4. Johannes M. Bauer and Michael Latzer (eds.), *Handbook on the Economics of the Internet* (Cheltenham, UK: Edward Elgar, 2016), https://doi.org/10.4337/9780857939852.

5. Erik Brynjolfsson, Daniel Rock, and Chad Syverson, "AI and the Modern Productivity Paradox: A Clash of Expectations," Working Paper 24001, November 2017, DOI 10.3386/w24001, https://www.nber.org/papers/w24001.

6. Rosellen Downey, "Top of the List: Apple, Google, Facebook Get Most Patents Approved in Silicon Valley," *Silicon Valley Business Journal*, accessed August 31, 2021, https://www.bizjournals.com/sanjose/news/2020/07/27/top-of-the-list-patent-recipients.html.

7. Ibid.

8. Ibid.

9. R. Polk Wagner and Thomas Jeitschko, "Why Amazon's '1-Click' Ordering Was a Game Changer," Knowledge@Wharton, September 14, 2017, https://knowledge.wharton.upenn.edu/article/amazons-1-click-goes-off-patent/.

10. Rosemarie Ham Ziedonis, "Don't Fence Me In: Fragmented Markets for Technology and the Patent Acquisition Strategies of Firms," *Management Science* 50, no. 6 (2004): 804–820.

11. Kaya Yurieff, "Apple and Samsung Settle Their Epic Patent Infringement Battle," CNN Business, June 27, 2018, https://money.cnn.com/2018/06/27/technology/apple-samsung-patent-infringement-settlement/index.html.

12. David J. Teece, "Profiting from Technological Innovation: Implications for Integration, Collaboration, Licensing and Public Policy," *Research Policy* 15, no. 6 (1986): 285–305, ISSN 0048-7333, https://doi.org/10.1016/0048-7333(86)90027-2 (https://www.sciencedirect.com/science/article/pii/0048733386900272).

13. Michael Lenox, Scott Rockart, and Arie Lewin, "Interdependency, Competition, and the Distribution of Firm and Industry Profits," *Management Science* 52 (2006): 757–772, 10.1287/musc.1050.0495.

14. Erik Brynjolfsson, Wang Jin, and Kristina McElheran, "The Power of Prediction: Predictive Analytics, Workplace Complements, and Heterogeneous Firm Performance," accessed August 31, 2021, https://www.hbs.edu/faculty/Shared%20Documents/conferences/strategy-science-2021/16_Kristina%20McElheran_The%20Power%20of%20Prediction%20Predictive%20Analytics,%20

Workplace%20Complements,%20and%20Heterogeneous%20Firm%20Performance.pdf.

15. Ibid.

16. David Teece, Gary Pisano, and Amy Shuen "Dynamic Capabilities and Strategic Management," *Strategic Management Journal* 18, no. 7 (1997): 509–533.

17. Public Broadcast System, "Bell Labs," ScienCentral Inc., 1999, accessed September 2, 2021, https://www.pbs.org/transistor/background1/corgs/bellabs.html.

18. Ingrid Lunden, "IBM Leads US Patent List for 2020 as Total Numbers Decline 1% in Pandemic Year to 352,000," TechCrunch, January 12, 2021, https://techcrunch.com/2021/01/12/ibm-leads-u-s-patent-list-for-2020-as-total-numbers-decline-1-in-pandemic-year-to-352000/.

19. Henry William Chesbrough, *Open Innovation: The New Imperative for Creating and Profiting from Technology* (Boston: Harvard Business Review Press, September 1, 2006), xxiv.

20. Gary Dushnitsky and Michael J. Lenox, "When Does Corporate Venture Capital Investment Create Firm Value?" *Journal of Business Venturing* 21 (2006): 753–772.

21. Michael Wayland, "GM-Backed Cruise to Buy Self-Driving Start-Up Voyage," CNBC, updated March 15, 2021, https://www.cnbc.com/2021/03/15/gm-backed-cruise-to-buy-self-driving-start-up-voyage.html.

22. Wesley M. Cohen and Daniel A. Levinthal, "Absorptive Capacity: A New Perspective on Learning and Innovation," *Administrative Science Quarterly* 35, no. 1, Special Issue: Technology, Organizations, and Innovation (March 1990): 128–152; 128.

23. Tiernan Ray, "MIT and Tsinghua Scholars Use DeepMind's AlphaFold Approach to Boost COVID-19 Antibodies," ZD Net, March 20, 2022, https://www.zdnet.com/article/mit-and-tsinghua-scholars-use-deepminds-alphafold-approach-to-boost-covid-19-antibodies/.

24. Clive Thompson, "The Long Game," *Technology Review* 125, no. 2 (2022): 78.

25. William M. Blair, "President Draws Planning Moral: Recalls Army Days to Show Value of Preparedness in Time of Crisis," *New York Times*, November 15, 1957, quote page 4, column 3, https://quoteinvestigator.com/2017/11/18/planning/#note-17261-6.

26. Kenneth Carrig and Scott Snell, *Strategic Execution: Driving Breakthrough Performance in Business* (Stanford, CA: Stanford Business Books, 2019).

Chapter 5

1. Grandview Research, "Cyber Security Market Size, Share & Trends Anal-

ysis Report by Component, by Security Type, by Solution, by Services, by Deployment, by Organization, by Application, by Region, and Segment Forecasts, 2021–2028 (Report Overview)," accessed September 6, 2021, https://www.grandviewresearch.com/industry-analysis/cyber-security-market.

2. "What Is Cybersecurity?" Cisco, accessed September 6, 2021, https://www.cisco.com/c/en/us/products/security/what-is-cybersecurity.html.

3. Alex Scroxton, "Average Ransomware Cost Triples, Says Report," ComputerWeekly.com, March 17, 2021, https://www.computerweekly.com/news/252498029/Average-ransomware-cost-triples-says-report.

4. Jake Frankenfield, "Data Analytics," Investopedia, accessed September 6, 2021, https://www.investopedia.com/terms/d/data-analytics.asp.

5. Alex Woodie, "Why Data Science Is Still a Top Job," Datanami, November 16, 2020, https://www.datanami.com/2020/11/16/why-data-science-is-still-a-top-job/.

6. Ibid.

7. Charles O'Reilly and Michael Tushman, *Lead and Disrupt: How to Solve the Innovator's Dilemma* (Stanford, CA: Stanford Business Books, 2016), 10.

8. Jeanne Liedtka and Tim Ogilvie, *Designing for Growth: A Design Thinking Tool Kit for Managers* (New York: Columbia Business School Publishing, 2011).

9. Ed Hess and Katherine Ludwig, *Humility Is the New Smart: Rethinking Human Excellence in the Smart Machine Age* (n.c.: Dreamscape Media, 2017).

10. Ibid., 5.

Chapter 6

1. Kashmir Hill, "How Target Figured Out a Teen Girl Was Pregnant Before Her Father Did," *Forbes*, February 16, 2012, https://www.forbes.com/sites/kashmirhill/2012/02/16/how-target-figured-out-a-teen-girl-was-pregnant-before-her-father-did/?sh=7b6025ec6668.

2. Anthony Cuthbertson, "Amazon Admits Employees Listen to Alexa Conversations," *The Independent*, April 11, 2019, https://www.independent.co.uk/life-style/gadgets-and-tech/news/amazon-alexa-echo-listening-spy-security-a8865056.html.

3. John Laidler, "High Tech Is Watching You," *The Harvard Gazette* (March 4, 2019), https://news.harvard.edu/gazette/story/2019/03/harvard-professor-says-surveillance-capitalism-is-undermining-democracy/.

4. Reuters, "Target Settles 2013 Hacked Customer Data Breach for $18.5 Million," NBC News, May 24, 2017, https://www.nbcnews.com/business/business-news/target-settles-2013-hacked-customer-data-breach-18-5-million-n764031.

5. Sam Varghese, "Ransomware Gangs Extracted US$81B in Ransoms in 2020 Sec Firm Claims," IT Wire, April 28, 2021, https://itwire.com/security/ran-

somware-gangs-extracted-us$18b-in-ransoms-in-2020,-sec-firm-claims.html.

6. Nathanial Lee, "Here's How Much Ransomware Attacks Are Costing the American Economy," CNBC, June 10, 2021, https://www.cnbc.com/2021/06/10/heres-how-much-ransomware-attacks-are-costing-the-american-economy.html.

7. David Sanger and Nicole Perlroth, "FBI Identifies Group Behind Pipeline Hack," *New York Times*, May 10, 2021.

8. Dave Davies, "Facial Recognition and Beyond: Journalist Ventures Inside China's 'Surveillance State,'" NPR, *Fresh Air*, January 5, 2021, https://www.npr.org/2021/01/05/953515627/facial-recognition-and-beyond-journalist-ventures-inside-chinas-surveillance-sta.

9. Paul Mozur, Muyi Xiao, and John Liu, "An Invisible Cage: How China Is Policing the Future," *New York Times*, June 25, 2022.

10. Victor Tangermann, "Elon Musk's Question for Super-Smart AI: What's Outside the Simulation?" Futurism, April 16, 2019, https://futurism.com/elon-musk-smart-ai-simulation.

11. Stuart Russell, *Human Compatible: Artificial Intelligence and the Problem of Control* (New York: Viking, 2019).

12. Statista, "Global Size of the Market for Industrial Robots Between 2019 and 2027," March 17, 2021, www.statista.com.

13. David Rotman, "How Technology Is Destroying Jobs," *MIT Technology Review* (June 12, 2013).

14. Ben Gilbert and Yuqing Liu, "This Chart Shows How the Salesforce Acquisition of Slack for $27.7 Billion Stacks Up Against Tech's Largest Deals Ever," *Business Insider* (December 2, 2020).

15. James Bressen, "Big Technology Is Slowing Innovation," *Technology Review* 125, no. 2 (2022): 80.

16. Ben Wolford, "What Is GDPR, the EU's New Data Protection Law?" GDPR.EU, accessed September 30, 2021, https://gdpr.eu/what-is-gdpr/.

17. California Consumer Privacy Act (CCPA), State of California Department of Justice, Privacy, accessed September 30, 2021, https://oag.ca.gov/privacy/ccpa.

18. *National Law Review* XII, no. 108 (April 18, 2022).

19. Devin Coldeway, "Who Regulates Social Media?" TechCrunch, October 19, 2020, https://techcrunch.com/2020/10/19/who-regulates-social-media/.

20. Caroline O'Donovan, "Uber and Lyft Spent Hundreds of Millions to Win Their Fight Over Workers' Rights. It Worked," BuzzFeed, November 21, 2020, https://www.buzzfeednews.com/article/carolineodonovan/uber-lyft-proposition-22-workers-rights.

21. Jonathan Vanian, "Why Companies Are Thinking Twice About Using Artificial Intelligence," *Fortune*, January 31, 2021, https://fortune.com/2021/01/31/ai-ethics-why-companies-are-thinking-twice-artificial-intelligence/.

22. R. Edward Freeman, Bidhan L. Parmar, and Kirsten Martin, *The Power of And: Responsible Business Without Trade-Offs*, (New York: Columbia Business School Publishing, 2020).

23. David P. Baron, "Disruptive Entrepreneurship and Dual Purpose Strategies: The Case of Uber," *Strategy Science* 3, no. 2 (June 19, 2018), https://pubsonline.informs.org/doi/10.1287/stsc.2018.0059.

24. Thomas Lyon, Magali A. Delmas, John W. Maxwell, Pratima Bansal, Mireille Chiroleu-Assouline, Patricia Crifo, Rodolphe Durand, Jean-Pascal Gond, Andrew King, Michael Lenox, Michael W. Toffel, David Vogel, and Frank Wijen, "CSR Needs CPR: Corporate Sustainability and Politics," *California Management Review* 60, no. 4 (Summer 2018): 5–24.

Chapter 7

1. "There's a .00006% Chance of Building a Billion Dollar Company: How This Man Did It," *First Round Review*, accessed September 30, 2021, https://review.firstround.com/Theres-a-00006-Chance-of-Building-a-Billion-Dollar-Company-How-This-Man-Did-It.

2. "The Climate Corporation," Khosla Ventures, accessed November 2016, http://www.khoslaventures.com/portfolio/climate-corporation; Bruce Upbin, "Monsanto Buys Climate Corp for $930 Million," *Forbes*, October 2, 2013, www.forbes.com/sites/bruceupbin/2013/10/02/monsanto-buys-climate-corp-for-930-million/#6e3557265ae1.

3. Michael Lenox and Rebecca Duff, *The Decarbonization Imperative: Transforming the Global Economy by 2050* (Stanford, CA: Stanford University Press, 2021).

4. Archie Mayani, "Health Care's Digital Transformation: Three Trends to Watch For," *Forbes*, July 26, 2021, https://www.forbes.com/sites/forbestechcouncil/2021/07/26/health-cares-digital-transformation-three-trends-to-watch-for/?sh=40639e3e5e8d.

5. Jacob Kastrenakes, "Beeple Sold an NFT for $69 Million," The Verge. March 11, 2021, https://www.theverge.com/2021/3/11/22325054/beeple-christies-nft-sale-cost-everydays-69-million.

6. Renuka Tahelyani, "Top 11 Most Expensive Bored Ape Yacht Club NFTS," *The Crypto Times*, March 30, 2022.

7. Andrew Hayward, "The Biggest Celebrity NFT Owners in the Bored Ape Yacht Club," Decrypt, March 27, 2022, https://decrypt.co/86135/biggest-celebrity-nft-owners-bored-ape-yacht-club.

8. Emma Boyles, "Second Life Realtor Makes $1 Million," Gamespot, November 27, 2006, https://www.gamespot.com/articles/second-life-realtor-makes-1-million/1100-6162315/.

9. David Yaffe-Bellany, Erin Griffith, and Ephrat Livni, "Cryptocurrencies Melt Down in a Perfect Storm of Fear and Panic," *New York Times*, May 12, 2022.

10. Bartu Kaleagasi, "A New AI Can Write Music as Well as a Human Composer," Futurism, March 9, 2017, https://futurism.com/a-new-ai-can-write-music-as-well-as-a-human-composer.

11. Robert Strohmeyer, "The 7 Worst Tech Predictions of All Time," PC-World, December 31, 2008, https://www.pcworld.com/article/532605/worst_tech_predictions.html.

Index

Page numbers in *italics* indicate figures.

Facial recognition systems, 11, 111,
    114–16
Facilitate strategies, 35
Fermi, Enrico, 115
Fintech companies, 51, 61, 143
First-mover advantages, 75–76, *76*
5G networks, 4, 7
Fixed costs, 26, 32
Fourth Industrial Revolution, 11
Freeman, Ed, 123
Freemium model, 9, 35, 58, 67, *68*, 132,
    143
Freemium models, 9, 35, *57*, 58, 67, 132
Friedberg, David, 128
Fuji, diversification by, 3
Fundamental Principal of Business
    Strategy, 73

Gamification, 55
Gaming platforms, 30
Gans, Joshua, 12, 99
Gap analysis, 65
General Data Protection Regulation
    (European Union), 119
General Electric (GE), 70–71, 97
General Motors (GM), 42, 46, 81–82, 97
General purpose technologies, 25, 28,
    29, 143
Global warming. *See* Climate change
Goldfarb, Avi, 12, 99
Google: AB testing by, 55;
    advertisements on, 57; Android
    OS, 24, 30, 31, 36, 42, 57, 58; in
    autonomous vehicle sector, 32, 46;
    competitive positioning by, 61; data
    collection by, 33, 42, 112–13; Gmail,
    56, 57; horizontal diversification

by, 31–32, 119; Maps, 38, 52, 56, 57;
    network externalities and, 58; patent
    portfolio of, 74, 75; predecessors to,
    42, 75; predictive analytics used by,
    79; primacy of data for, 2; search
    technologies, 6, 11, 25, 31
Graef, Ailin, 136

Hacking, 95, 114
Hawking, Stephen, 115
Helmer, Hamilton, 61, 64
Hess, Ed, 12, 106–7
Horizontal diversification, 31–32,
    118–19, 143
Hounsfield, Godfrey, 71
Humility, 62, 106
Hypercompetition, 42, 75, 80

IBM: digital transformations by, 2, 3, 64,
    71; Maersk partnership with, 7, 98;
    open strategy pursued by, 37–38, 77;
    Watson Research Center, 80
Immelt, Jeff, 70
Industrial Revolution, 11, 28, 117
Infrastructure. *See* Digital infrastructure
Innovation: acceleration of, 2;
    appropriability and, 71, 77, 80–84;
    architectural, 41; automation
    of, 82; in business models, 30;
    commercialization of, 77; data
    analytics and, 11; design thinking
    and, 103; disruptive, 80, 83, 143;
    downstream, 28, 30, 31; enabling
    technologies and, 28, 30; open
    model of, 81–82, 102; radical, 41,
    79; user-led, 55, 84; value chain
    deconstruction and, 50–51; value